*A History of the
Photographic Lens*

A History of the Photographic Lens

Rudolf Kingslake

The Institute of Optics
University of Rochester
Rochester, New York

ACADEMIC PRESS, INC.

Harcourt Brace Jovanovich, Publishers

Boston San Diego New York
Berkeley London Sydney
Tokyo Toronto

ACADEMIC PRESS, INC.
1250 Sixth Avenue, San Diego, CA 92101

United Kingdom Edition published by
ACADEMIC PRESS INC. (LONDON) LTD.
24–28 Oval Road, London NW1 7DX

Library of Congress Cataloging-in-Publication Data
Kingslake, Rudolf.
    A history of the photographic lens/Rudolf Kingslake.
       p.   cm.
    ISBN 0-12-408640-3
    1. Lenses, Photographic. I. Title.
  TR270.K49   1989
  771.3′52′09 — dc19   88-34299
                                                      CIP

Printed in the United States of America
89 90 91 92   9 8 7 6 5 4 3 2 1

CONTENTS

PREFACE

My interest in lens design began while I was still a schoolboy on reading a book of my father's entitled *Photographic Lenses: A Simple Treatise*, by C. Beck and H. Andrews, published about 1907. In this book were several sectional diagrams of lenses with graphs of astigmatism and field curvature, which I did not fully understand. However, the seed was sown, and when the time came to go to college I immediately chose the Imperial College of Science and Technology, in London, where there was a course of lectures on lens design given by Professor A.E. Conrady.

Later, in 1929, after holding a couple of smaller jobs in England, I was offered an assistant professorship at the newly established Institute of Applied Optics at the University of Rochester, which I immediately accepted. After another eight years an opportunity for further advancement came when I was chosen to succeed C.W. Frederick as head of the lens design department at the Eastman Kodak Company, from which I retired in 1968, 31 years later.

So, the history and design of photographic lenses has been my major interest for nearly seventy years. It is a fascinating subject, involving inevitably the greatest pioneers of lens design, Petzval, Dallmeyer, Rudolph, Taylor, Lee, Bertele, and more recently Naumann, plus numerous living designers that I have not included here. It is, of course, impossible to mention all of the thousands of different lenses that have been developed during the 150 years since photography was introduced, nor is it possible to include the hundreds of men who have worked laboriously to improve photographic objectives, but I have tried to summarize the major developments in this field, with a few minor ones as well.

It is hoped that the story given here will be of interest to those photographers who desire to know more about their equipment and how it came to be developed.

Rudolf Kingslake
June 1989

CHAPTER 1

Introduction

The science and art of photography have been thoroughly documented throughout its history. Many writers have been attracted to this subject, and innumerable photographic books and journals have flourished since the announcement of Daguerre's invention in 1839. The photographic lens has by no means been neglected; several books on this subject have been particularly useful to the historian (see Section VII). Nevertheless, it remains true that the lens is generally the most expensive and least understood part of any camera.

Of course, ordinary photographers have no difficulty in taking excellent photographs without knowing anything about the construction of their equipment. Most workers distinguish completely between the artistic and technological aspects of the subject. A similar situation exists in many other aspects of modern life. For example, how many users are familiar with the internal construction of their radio, typewriter, watch, TV set, calculator, or even ordinary motor engine? Yet these things are used all the time by people who know nothing about them, and in many cases they couldn't care less. Nevertheless, there are some who are anxious to know more about their tools and instruments. It is to these people, and to the growing number of collectors of cameras and other photographica, that this book is directed.

I. LENS DESIGN

Lens design is far from a simple matter. The function of a lens is to take a bundle of rays emanating from a single object point and bend each ray by just the right amount in just the right direction to make all the rays meet again exactly at the corresponding point in the image. As if this were not difficult enough, the image point in question must lie on a flat plane (the surface of the plate or film), and the magnification (i.e., the ratio of the image height to the object height) must be constant over the entire field of view. In addition, the rays in all colors must be made to meet together, in spite of the fact that the refractive index of all transparent substances varies from one color to another.

If a young engineer were presented with this problem in its complete form, he or she would at once declare it to be hopelessly difficult, if not insoluble. Indeed, most of the opticians existing when photography was introduced had no idea how to design a lens that would cover a wide flat field, as this problem had never arisen. Yet we know that excellent photographic lenses have been designed and manufactured for over 140 years. The explanation is, of course, that lens designers approached the problem in steps. The earliest objectives were single elements in which the shape and stop position of the lens were chosen in such a way as to optimize the image quality over the field. Then it was found necessary to achromatize the system, and cemented doublets were introduced. Next, distortion was found to be a problem, so a pair of identical elements were mounted on each side of a central stop to remove it. Astigmatism was found to be the principal remaining defect and, by 1890, means had been found to eliminate it, and so on.

Fortunately, there are several laws governing lens behavior that the designer soon learns and that are of inestimable value to him in solving what would otherwise be an impossible problem. For instance, he soon finds that symmetry about a central stop has the effect of automatically correcting three difficult aberrations (see Chapter 4, I), for which reason symmetrical lenses remained popular for nearly a hundred years. He also finds that the astigmatism in an image is controlled by a theorem named after Petzval, which will be discussed in Section B. With these and other well-known optical laws before him, a designer finds that the determination of lens structure becomes much simpler, and in due course he can produce a design that promises to be good enough to justify making up a sample for testing. Actually, very few lenses are found to be perfect at this first trial, and a continuing series of changes and new developments follow, resulting in either a satisfactory design or a decision to abandon the attempt and try something different.

In retrospect, it is hard to understand why the development of a good camera lens was such a slow process, especially between the years 1840 and 1890. The mechanical skill of some of the earlier lens manufacturers was remarkable, and many historical lenses show evidence of excellent design and construction. (Other lenses, unfortunately, are sometimes so bad that one wonders how they could have been used at all.) We are driven to the conclusion that the lens designers of the day were slow to learn and apply the theory of lenses as developed by Gauss, Petzval, von Seidel, and others. Opticians were far too willing to make empirical trials and put together a series of lens elements, hoping that a miracle would happen and that their system would turn out to be better than those currently available. There were exceptions, of course. Petzval's Portrait lens was designed on paper before fabrication was commenced, using theoretical formulae that he developed himself, while Steinheil's Aplanat of 1866 was aided by the theoretical work of von Seidel. Even these successes did not convince the optical world of the need for a good theoretical background. H. Dennis Taylor (1862–1943), the designer of the Cooke Triplet lens, made only partial use of theory, and he would often adjust a new lens in the workshop until the best image was obtained. Today a new design can be analyzed completely by computer, and it is no longer necessary to construct a sample to determine whether the lens is good enough for its intended application.

We must realize, however, that some interesting lens designs were once considered unsatisfactory because no one was willing to devote the time and patience necessary to explore the possibilities of that particular type. Some of these have returned to popularity now that electronic computers enable us to try out a multitude of possible designs in a very short time.

It is interesting to note that the earliest lens to be properly designed, the simple Wollaston Landscape lens of 1812, is still being manufactured in enormous numbers for low-cost cameras; the Petzval Portrait lens of 1840 was made and used regularly until about 1920; the Rapid Rectilinear lens of 1866 was fitted almost universally to all the better cameras for over sixty years; and the Cooke Triplet, after a few years of hesitation, has become the universal design for all cameras and projectors where an aperture of $f/3.5$ or less is adequate. Many other types of construction have come and gone, while some, such as the telephoto, have returned to popularity after a lapse of many years. Innumerable experimental lenses are constantly being developed, but most of these show no advantage over existing types, and they are soon abandoned and forgotten.

A. Lens Aberrations

During the 1850s, L. von Seidel (1821–1896) investigated the formation of an image by a lens and identified seven so-called primary aberrations that affect the definition. These aberrations are independent of one
another, and correcting one does not imply that any of the others will be
corrected. Indeed, it is generally found that we require one degree of
freedom in a lens (e.g., the curvature of a surface, the thickness of an
element, the airspace between two elements, or the position of the stop) to
correct one aberration. Hence, if six aberrations are to be corrected there
must be at least six degrees of freedom available to the designer. Actually,
there must be one more degree of freedom with which to hold the focal
length. This is because a lens can be constructed to any desired focal length
by merely scaling all the linear dimensions; consequently, a degree of
freedom is really the ratio of one variable parameter to another, so there
must be one more available parameter than the number of aberrations to
be corrected.

Most good lenses possess more than the minimum number of degrees of
freedom. This helps the designer, because there are several important
aberrations that do not have any primary value, and even the well-known
Seidel aberrations have higher-order components that must be controlled
separately. In the simpler types of lens, the designer has to accept whatever
higher-order aberrations happen to appear, as he has no control over
them. Of course, he can try the effect of a change of glass or he can even
resort to the use of one or more aspheric surfaces, although this latter
possibility is likely to prove very expensive to manufacture.

B. The Petzval Sum

The design of any photographic lens is dominated by a certain mathe-
matical expression known as the Petzval sum. This sum is related to the
basic field curvature that we should expect to find in any lens, from very
simple considerations. In Figure 1.1, it is clear that the oblique object
distance b is greater than the axial object distance a; hence, the oblique
image distance b' should be less than the axial image distance a', leading at
once to a curved image for a plane object.

Joseph Petzval (1807–1891), in 1839, reduced this problem to mathe-
matical terms and showed that if all other aberrations are absent, the
curvature of the image of a flat object will be given by the sum $\Sigma \, \phi/nn'$, ϕ
being the power of a lens surface, given by $(n' - n)/r$ where r is the radius
of curvature of the surface and n and n' are the refractive indices of the

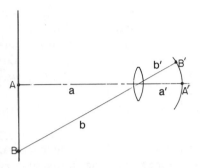

Figure 1.1. The reason why a simple lens has a curved field.

media on the two sides of that particular surface. This sum differs from all other aberrations in that it does not depend on the aperture or field of the lens nor on the stop position, airspaces, thickness of the elements, or conjugate distances. Indeed, some workers do not regard it as an aberration at all but merely an inherent property of the lens along with its focal length and other basic data.

If the system consists of a series of separated thin elements, the Petzval sum becomes merely $\Sigma\ \phi/n$, where ϕ is the lens power and n is the refractive index of the material from which it is made. The Petzval sum of a thin lens of ordinary crown glass is about 0.66 for a focal length of unity or 0.066 for a focal length of 10.

In practically every lens made prior to about 1885, this natural inward field curvature was deliberately offset by the introduction of some degree of overcorrected astigmatism, which served to flatten the field artificially by opposing the inward field caused by the Petzval sum. Thus, the designers of these early lenses traded astigmatism for field curvature, or as Piazzi Smyth remarked in 1874[1], they "relieved us of a blunder by substituting a sin." At apertures as low as $f/15$ the astigmatism so introduced did not become serious until about 25° from the lens axis, but in a portrait lens with an aperture of $f/3.5$ the astigmatism was liable to become serious at perhaps 10° out. In all the early lenses it was the increasing astigmatism that set a limit to the useful field, leading to the well-established idea that a high aperture and a wide angular field were incompatible.

In some early portrait lenses a considerable amount of inward field curvature was deliberately left in the lens by the use of less overcorrected

[1] *Brit. J. Phot. Almanac*, 1874, p. 44.

astigmatism. This improved the sharpness of definition out in the field, but projecting the flat film plane out into the object space led to a curved object field. For this reason, a row of people would be seated on a curve rather than in a straight row. Photographs were often taken in the corner of a room or looking down a street, so that the objects of principal interest would be sharply imaged on the plate.

There are three principal ways by which the Petzval sum of a lens can be reduced:

(a)   By the use of a thick meniscus construction, the two outer lens surfaces having about the same radius of curvature.

(b)   By the use of a series of positive and negative lens elements widely spaced. Achromatizing such a system requires that the negative elements be considerably strengthened, which immediately reduces the Petzval sum.

(c)   By the use of a crown glass of high refractive index combined with a flint glass of lower refractive index.

All these methods have been used in lenses, but oddly it was the third that was tried first. By 1886 Abbe and Schott in Jena had developed barium crown glasses having just the required property to reduce the Petzval sum, and these glasses were immediately adopted by Schroeder in 1888 in his Ross Concentric lens.[2] Soon the other methods were tried by almost every designer, resulting in the development of numerous anastigmat lenses. A few nonanastigmats are still used, provided that the aperture or field are low enough so that the inevitable astigmatism is not serious or where the field is deliberately made inward-curving so that the image lies on the Petzval surface and the astigmatism automatically disappears.

C. Remounting a Lens

If an occasion arises to remount the two halves of a photographic objective in a new barrel or shutter, it is essential to ensure that the optical axes of the two halves are coincident and that there is the correct separation between the two components. Some lenses are highly sensitive to the space between the halves, and an error of as little as 0.5% of the focal length can often be quite serious, while in other lenses the central airspace may be insensitive. For instance, in a lens of the Dagor type, a small error in the separation by as little as 0.4% has a considerable effect on the field curvature, and also a lesser effect on the coma, an increase in the airspace

---

[2] U.S. Pat. 404,506 (1888).

leading to an inward-curving field. On the other hand, in a lens of the Rapid Rectilinear type, an error as large as 1.0% has only a negligible effect on the field curvature and almost no effect on the coma, an increase in the airspace leading to a backward-curving field. In general, it is usually found that the better the lens the greater is the effect of an error in the central airspace.

II. A BRIEF HISTORICAL SURVEY

Any attempt to develop a strictly chronological approach to the history of the photographic objective is invariably confused by a hopeless mass of crosscurrents. It would be easier if each type of lens had been invented, developed, perfected, and then abandoned in a limited period of time, after which another type had appeared and been similarly treated. Unfortunately for the historian, but fortunately for the working photographer, the lenses available at any one time cover a wide range of constructional types, some being suitable for use at high apertures, some covering a wide angular field, some being usable over a broad range of magnifications, some being simple and cheap, while some are complicated and expensive. It is, in fact, only recently that a high aperture and a wide angular field have become possible simultaneously in the same lens, and it is only in the last forty years that lenses with a variable focal length have become available. Consequently, in this book it has been decided to follow the rise and decline of each type separately, starting with its first introduction and continuing its development to the present day, or until it was abandoned.

A. 1840 to 1866

Photography was invented by L. J. M. Daguerre (1789–1851) in 1839, and the first lens to be used on a camera was the achromatic landscape lens of C. Chevalier (1804–1859). The aperture of this lens was only $f/15$, and the need for a much faster lens was immediately obvious. This requirement was soon met by J. M. Petzval in his famous Portrait lens of 1840. A few high-aperture objectives were developed by other workers but they were not good enough for practical portraiture.

Partly because of the slowness of the original daguerreotype process, many photographers were in the habit of photographing inanimate objects such as fine buildings; the appearance of barrel distortion in landscape lenses became evident by the curvature of straight lines in the outer parts of the field. So a demand arose for distortionless lenses, and a number of symmetrical systems were developed empirically by such workers as Cundell, Ross, Davidson, Sutton, and others. The first symmetrical objectives to be properly designed were the 1859 Panoramic lens of T. Sutton (1819–1875), the 1865 Periskop of C. A. Steinheil (1801–1870), and the 1860 Globe lens of Harrison and Schnitzer in New York.

A few unsymmetrical distortionless lenses appeared on the market. In 1859, when the cry for distortionless lenses was at its height, the Petzval Orthoskop appeared. This lens was hailed as the solution for all the photographer's problems, but careful investigation showed that it was not perfectly distortionless and the field was somewhat curved, so the new lens was soon forgotten.

B. 1866 to 1890

In 1866 the Dallmeyer Rapid Rectilinear and the identical Steinheil Aplanat were produced. At that time there were four types of lens available to the working photographer, namely, the Landscape lens, the Portrait lens, the wide-angle Globe lens, and the intermediate Rapid Rectilinear, which together met most of their needs. However, all these lenses suffered from overcorrected astigmatism in the outer parts of the field, which was serious in lenses of high aperture but scarcely significant in lenses of very low aperture.

C. 1890 to 1914

The correction of astigmatism and the development of the Anastigmat lens of 1890 were made possible by the introduction of barium crown glasses by E. Abbe (1840–1905) and O. Schott (1851–1935) of the Zeiss Company, in 1885. Now for the first time it was possible to eliminate the astigmatism that had bedevilled all previous lenses, leading to the design of such well-known objectives as the Tessar, Dagor, and many others. Nevertheless, old favorites such as the Petzval Portrait and Rapid Rectilinear lenses took a long time to disappear. They were still being made and used extensively by 1930. The year 1890 also saw the introduction of telephoto lenses by Dallmeyer and others and, on the whole, it was a period of steady improvement in lenses of many new types.

D. The Interwar Period, 1918 to 1940

By 1920 lenses of $f/2$ aperture were becoming available, and the period around 1930 was particularly active in the development of new lenses. Reversed telephotos were appearing, at first for close-up projection and then for use in the Technicolor three-strip camera, where a long back focus clearance was needed to house the beam-splitting prism. Some primitive zoom lenses were being developed for professional 35 mm movie cameras, and the use of anamorphic compression in motion pictures was being proposed by Chrétien. During this period, 16 mm and 8 mm movies were being developed for the amateur photographer, requiring the design of many new lenses. Toward the end of the period, some very large lenses were being developed for use in aerial cameras.

E. The Post World War II Period

As soon as World War II ended, there was a great rush of pent-up activity in the design of new photographic lenses. Practical zoom lenses for 35 mm motion pictures soon appeared, and these were followed by lenses for the new field of television. The CinemaScope system of 1952 gave the cinema a much-needed boost; this involved the use of anamorphic compression in both the camera and the projector. New lenses of high quality were required for the new film formats of the 126, 110, and Disc cameras.

Starting in the early 1950s, Japanese camera and lens manufacturers began to take over the amateur camera market and soon outstripped European countries in the low cost and high quality of their products. Eventually, even the German manufacturers could no longer compete, and today Japan has a virtual monopoly in the 35 mm still-camera market.

As an example of Japanese enterprise, we may consider the offering of a single company, Nikon, which currently makes over fifty lenses for 35 mm SLR cameras. They have three fish-eye lenses from 6 mm to 16 mm focal length; eleven reversed telephoto wide-angle lenses from 13 mm to 35 mm; three normal lenses of 50 mm focal length; twenty-three telephotos from 80 mm to 1200 mm; three mirror lenses from 500 mm to 2000 mm; and eleven zoom lenses ranging from 28–85 mm to 180–600 mm. Several other companies are manufacturing similar lines of first-quality lenses, all reasonably priced and readily available in this country.

These developments have been made possible by the introduction of the high-speed digital computer (see Section VC). Other contributing factors are efficient antireflection coatings and the introduction of new types of high-index optical glass.

III. LENS MARKINGS

At first very few lenses carried any indication of the maker or the structure of the lens. There were exceptions, of course, but often the lens would be housed in a polished brass barrel and that was it. The user had to fit the best lens he could find to his camera, and most photographers were reasonably successful in doing so.

By the middle of last century, lens names began to appear. These were proprietary and indicated both the manufacturer and the type. Owners of a Tessar or a Heliar would be very proud of having such a fine lens. Eventually, however, the same name might be used for various types of construction. Today, lens names have almost entirely disappeared, with only the maker's name and the lens data being engraved on the mount.

In a few cases, importers have marked their names on the lenses that they sell; the actual maker is unknown to the purchaser. Sometimes a patent number or the date of issue of a patent has been engraved on the lens mount. Although this was presumably done to discourage copying, it actually serves as a useful indication of the lens structure.

Many lenses carry a serial number alongside the other data. This can provide an indication of the year of manufacture, although it is sometimes coded in such a way that the owner cannot interpret the number; more often, it serves merely as an identification label for insurance purposes.

A. Aperture Control Devices

In the days of the daguerreotype, obviously no means for reducing the lens aperture was required, as photographers needed all the light they could get. However, with the introduction of the wet collodion process by F. Scott Archer (1813–1857) in 1851, the emulsion speed was so great that some means had to be found for reducing either the lens aperture or the exposure time. The latter possibility had to wait for the invention of mechanical shutters in the 1870s, but a simple means for reducing the lens aperture was introduced by John Waterhouse in 1858.[3]

The Waterhouse stop consisted of a flat piece of metal having a circular or other hole to form the actual aperture, which was slipped into a slot at the correct place in the lens barrel. Many early lenses were accompanied by a neat leather case holding a set of Waterhouse stops to be used with that particular lens.

Some of the early Jamin-Darlot lenses were equipped with three swing-out stops on levers so that any stop could be swung into position as desired. Another common arrangement was a wheel of stops pivoted at one side of the lens barrel, which could be rotated to bring the desired aperture into position.

---

[3] *Phot. J.* **4**, 258 (1857–58).

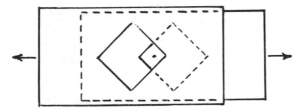

Figure 1.2. The Noton cat-eye diaphragm.

The commonest means for aperture control is the familiar iris diaphragm. It is not known who invented this extremely ingenious mechanism, but it was apparently known early in the last century. The mechanism consists of a set of identical arcuate leaves mounted around the lens aperture, each leaf having short studs at the two ends, one stud being at the front and the other at the back of the leaf. All the studs above the leaves fit into holes spaced around a metal ring, while the studs below the leaves fit into radial slots in a second ring. When one ring is rotated relative to the other, all the leaves move inward, creating an aperture of reduced diameter. If only three leaves are used, the resulting aperture is a curvilinear triangle, but if a large number of leaves are used, the result is a good approximation to a circle. Sometimes bright lights in a night photo exhibit radial streaks on the negative. These are due to diffraction at the edges of the diaphragm leaves, there being the same number of streaks as there are diaphragm leaves.

· In 1858 C. C. Harrison of New York patented a novel type of iris diaphragm in which both studs were at the same end of the leaf, the length of the leaf being slightly greater than the semidiameter of the aperture.[4] This form of iris was little used at the time, but when iris shutters were introduced in the 1880s, the shutter blades were invariably constructed on Harrison's principle, there being anywhere from three to twelve or more blades around the aperture. Today, when most cameras are equipped with some kind of automatic aperture control, Harrison's iris is the only kind that is capable of responding sufficiently rapidly to be useful. If Harrison were living today, he would have profited greatly from his invention.

Another type of aperture control that is occasionally used is the cat-eye diaphragm, originally proposed by M. Noton in 1856 (Fig. 1.2). Here two overlapping metal plates carry square openings, which together form a

---

[4] U.S. Pat. 21,470 (1858).

square aperture, its size being continuously variable by moving the two plates at equal rates in opposite directions.

B. Aperture Designation

Several systems of aperture designation have been introduced over the years. Some early lenses carried diaphragm markings that merely indicated the diameter of the iris opening in millimeters, leaving it to the user to interpret this in relation to the focal length of the lens and the required aperture.

About 1880 it began to be understood that the speed of a lens and also the depth of field are dependent upon the slope angle U' of the steepest ray emerging from the lens and proceeding to form the axial image. Lens speed is thus proportional to $\sin^2 U'$, the quantity $\sin U'$ being known as the numerical aperture of the lens. For photographic purposes, it soon became the custom to designate the lens aperture by the ratio of the focal length to the diameter of the entrance pupil, the so-called F-number of the lens. In Europe and Japan this ratio is indicated as a fraction, (e.g., $1:4.5$ if the focal length is 4.5 times the aperture diameter). In English-speaking countries the diameter of the aperture is usually expressed as a fraction of the focal length (e.g., $f/4.5$). The denominator, in this case 4.5, is called the F-number of the lens, often represented by the symbol N. Unfortunately, a high value of N indicates a small lens aperture and vice versa, which can be confusing.

The relation between the F-number and the numerical aperture is given by:

$$N = \frac{1}{2 \text{ N.A.}} = \frac{1}{2 \sin U'}$$

or

$$\sin U' = \frac{1}{2N}.$$

The first standard series of F-numbers was that proposed by Franz Stolze (1830–1910)[5]:

$f/1.1$, 1.6, 2.2, 3.3, 4.5, 6.3, 9, 12.5, 18, 25, 36, 50, 71, 100.

---

[5] C. Diserens, *Traité de Photographie: Vol. 1, Optique*, p. 326. Gauthier Villars, Paris, 1943.

The ratio of each number to the next is the square root of 2. For many years, most European lenses were marked according to this system. However, in recent years it has become customary to use a different series, which is now universal:

$$f/1, 1.4, 2, 2.8, 4, 5.6, 8, 11, 16, 22, 32, 45, 64, 90.$$

The maximum aperture of a lens can be anything, but the smaller apertures are invariably marked with one of these series.

In the past, a few attempts have been made to mark the lens aperture by the relative image illumination. Paul Rudolph of Zeiss proposed two such systems. In his first system the stop was marked by the approximate value of $10,000/N^2$. Thus, $f/9$ became 128 in his system, and so on. Later he divided these numbers by 4 so that $f/9$ became 32.

Around 1890 the opposite scheme was proposed, in which the aperture markings were intended to be proportional to the exposure. This was called the Uniform Scale system, abbreviated to U.S. Here U.S. $= N^2/16$, so that $f/16$ became U.S. 16, and $f/4$ was marked U.S. 1. For many years all of the lower-priced Kodak and other cameras were marked on this system.

C. Focusing Devices

Lenses intended to be used on a bellows camera needed no focusing arrangement, but they were often equipped with a rack and pinion to provide more sensitive focusing than was possible by using the heavy front board and rear supports of a large wooden camera. For smaller cameras, a helical focusing arrangement was often used.

The required movement of a lens from its infinity position when focusing on an object at a distance L from the lens is given by the formula

$$X = f^2/(L - f),$$

where f is the focal length. Thus, a lens of 4 inches focal length, when focused on an object 3 feet away, must be moved forward from its infinity position by a distance of $4^2/(36 - 4) = 0.5$ inch. This movement can be handled without difficulty, but a larger focusing movement becomes mechanically troublesome.

The quantity L in this formula is, strictly speaking, the distance of the object from the front nodal point of the lens. As the location of this point is generally unknown, particularly in the case of a long telephoto or zoom lens, a better procedure is to measure the distance S of the object from the film plane, which is often marked by the symbol ϕ on the top of the camera. The required lens movement X when the distance S is known is found from the quadratic equation

$$X^2 - X(S - 2f) + f^2 = 0.$$

About 1900 someone suggested focusing a lens by moving only the front element instead of the whole objective. The advantage of this procedure is that in a Triplet or a Tessar the power of the front element is about three times the power of the whole objective, so that the required focusing movement is only about one-ninth as great as for the lens as a whole. Furthermore, if only the front element is to be moved, the rest of the lens and the shutter can be rigidly fixed in position. The penalty of using this arrangement is that the aberration correction carefully built into the design is completely upset if the front airspace is altered. However, by over-correcting the spherical aberration at the infinity setting, where the lens will generally be used stopped down, the undercorrection at close distances will be reduced. Also, when the object is close to the camera, the image will be large, and some degree of spherical aberration can be tolerated. With the coming of miniature cameras, the need for front-element focusing has largely disappeared, although it is still used in some zoom lenses.

Recently a few lenses have been constructed that are focused by a movement of an internal rear element, the remainder of the system remaining fixed. This arrangement is particularly convenient for cameras equipped with some type of automatic focusing.

Even when the entire lens is moved for focusing, some changes in the aberration correction can be expected. One proposal to maintain good image correction with a very close object is to move one of the lens elements (a floating lens) at a different rate from the rest of the system. Such an objective is known as a macro lens.

IV. IDENTIFYING A GIVEN LENS

The various photographic museums around the world possess numerous examples of early lenses, but because many of these carry no engraving of any kind, the curator often has difficulty in identifying some particular lens or even determining who made it. Some lenses carry the maker's name but no indication of the type of construction. If the lens has a long barrel with rack-and-pinion focusing, the chances are that it is a Petzval portrait lens, but, of course, this conclusion is not always valid. In a few cases the maker's name and other data have been written by hand on the ground rim of a negative element in the lens.

If the lens can be readily disassembled into its separate elements, the type of construction can be quickly determined, but if cemented components are involved, identification becomes more difficult. One possible procedure is to remove the cemented unit from its mounting and hold it horizontally beneath a single light bulb near the ceiling. On slightly tilting the lens, the various reflected images of the light will move across it, and a little thought will often indicate whether the surface causing the reflection is convex or concave toward the observer. Uncoated glass-air surfaces yield bright images, while cemented surfaces form faint reflections. For example, one-half of a Rapid Rectilinear lens contains two external surfaces and one internal surface all facing the same way, so that when this lens is tilted the three reflections move in the same direction but at different rates. On the other hand, in the front component of a Petzval Portrait lens, the interface reflection and the reflections from the two outer surfaces move in opposite directions. In a cemented triplet such as one-half of a Dagor, the two cemented interfaces have opposite curvatures and sometimes with a little care it is possible to identify the lens structure. This process becomes much more difficult if the outer lens surfaces are coated.

In some cemented lenses there is a visible groove or step around the edge of the lens where the cemented interface emerges, but in some cases the edge of the intersection is barely visible even under a magnifier. If the positive element in a cemented combination runs out to a knife edge, it is virtually undetectable.

V. SOME RECENT DEVELOPMENTS

A. Aspheric Surfaces

The only kind of surface that can be conveniently and accurately generated on a piece of glass is a sphere or a plane. When several elements are assembled in a lens barrel, the centers of curvature of all the lens surfaces must be made to lie on a straight line called the lens axis. Failure to achieve this will result in a decentered lens, and good definition becomes impossible.

It has long been known that the use of a nonspherical surface of revolution on a lens would give the designer several additional degrees of freedom with which to work, but the practical problems involved in the generation of such surfaces on a piece of glass are horrendous. Such a surface has its own axis, of course, and this axis must be made to coincide with the axes of the other surfaces in the system. Nevertheless, in spite of these problems, some success is being achieved in the production of lenses having aspheric surfaces. Kodak, Corning, and other companies have recently had some success in molding glass to an accurately known aspheric surface[6]. Of course, it is just as easy to mold a plastic asphere as a spherical surface; the only problem is how to make the aspheric mold. One good mold can be used to make thousands of plastic aspheres, and this may well be the way in which low-cost aspheres will be made. Several recent patents have appeared for high-aperture television projection lenses in which two or more plastic elements have aspheric surfaces on one or both surfaces. Plastic biaspheres are being proposed for the high-aperture lenses used in video disc recorders. It has even been proposed to make a base lens of glass and cover it with a thin layer of epoxy in which an aspheric surface has been pressed.

B. Antireflection Coatings

An important limiting factor in all early lenses was the presence of interreflection of light between the various glass-air surfaces in the lens. This led to a general loss of light in the image, but the worst consequence was the formation of ghost images and flare spots in the picture. Lenses having two or four glass-air surfaces did not suffer unduly in this respect, and six glass-air surfaces were generally considered acceptable. Lenses such as the original Anastigmat and the Dagor were admired, in spite of their excessive zonal aberration, largely because they had only four reflecting surfaces. However, when eight-surface lenses began to appear, such as the Celor of Goerz and the Unar and Planar of Zeiss, some users became unhappy and maintained vigorously that the old landscape lenses used in simple box cameras yielded negatives with more contrast. Certainly no designer dared to go beyond that limit.

Around 1896 H. Dennis Taylor observed that some old lenses that had become tarnished by exposure to the atmosphere actually transmitted more light than a newly polished lens. He reasoned that the thin layer of tarnish on the lens surfaces had a lower refractive index than the body of the glass, leading to a reduction of loss by reflection and a consequent increase in the light transmission. In 1904 Taylor patented the use of acids or other chemicals to cause tarnish deliberately on the surfaces of a lens, to

---

[6] P. L. Ruben, "Design and Use of Mass-produced Aspheres at Kodak." *Appl. Opt.* **24**, 1682 (1985).

reduce reflectivity.[7] However, this process proved to be uncertain in operation, some of the lenses in a batch becoming tarnished quickly while others did not tarnish at all or only very slowly.

The whole subject was opened up in 1936 when A. Smakula of Zeiss invented the process of coating lens surfaces in a vacuum with a thin evaporated layer of low-index material.[8] A single layer of such material, originally calcium fluoride or magnesium fluoride, has the effect of reducing the reflectivity greatly at one wavelength in the middle of the spectrum, but less at the longer or shorter wavelengths at the ends of the spectrum. Later on, various plans were presented for the evaporation of several layers of differing materials on top of one another, by which means the reflectance could be reduced almost to the vanishing point throughout the entire spectrum.

The discovery of antireflection coating of lenses opened up the whole field of multielement optical systems, so that now almost any desired number of airspaced elements can be used without danger of ghosts and flare. Without such coating all the modern zoom and other complex lenses would be impossible. Of course, a layer of cement used to join two lens elements together cannot be coated, and such a surface may reflect a considerable amount of light. Today one occasionally sees a row of ghost images in television images when the sun is in or close to the picture, in spite of the efforts of the designer to avoid this possibility.

C. Computers

By far the most important recent advance in lens-design technology has been the advent of the digital computer. No matter how it is performed, lens design necessarily involves an enormous amount of numerical computation ("number crunching"). These computations consist mainly, but not entirely, in the tracing of numerous selected rays through a possible lens using eight-place trigonometry. In the old days this was performed by hand with the help of a table of logarithms, and the time required to trace a ray through a spherical refracting surface might run anywhere from two to five minutes, or even longer toward the end of the afternoon! Motor-driven mechanical desk calculators, starting from about 1930, were a help, and by using today's programmable electronic pocket calculators this time has been reduced to about five or six seconds per ray-surface and a great

---

[7] Brit. Pat. 29,561/04.
[8] Ger. Pat. 685,767 (1936).

deal less when a minicomputer or microcomputer is used. The largest computers are capable of tracing over 12,000 ray-surfaces per second.

At speeds such as this there is no sense in tracing just one ray, and many sophisticated programs have been written where enough rays are traced to provide an estimate of the anticipated performance of a lens, followed by a systematic series of changes in the lens structure, with an indication in each case as to whether the change is beneficial or otherwise. By repeated application of this procedure the design of a lens can be progressively improved until it is as good as it can possibly be. If it is still not good enough for the intended application, the designer makes a major change in the lens structure and repeats the optimization process until a satisfactory design is obtained. Hence, many modern lenses are an order of magnitude better than any previous lenses that had been designed by hand. It would take far too long to try out every possible change if all the computing had to be done by logarithms or a desk calculator.

D. Gradient Index Material

The latest development in lens construction is to use glass having a gradient of refractive index, either a radial gradient increasing out from the axis of the piece or a longitudinal gradient increasing along the axis. Some lenses have recently been designed using currently available techniques for producing a gradient of refractive index, the result being equivalent to an aspheric surface without the problems of making such a surface. This subject is still in its infancy, but the prospects are good.

VI. LENS PATENTS

A patent is a legal monopoly granted by the government for a limited period of time; in the United States it is seventeen years from the date of issue of the patent. Possession of a patent gives patentees the right to prevent others from making, selling, or using their devices. A patent is *not* a certificate of invention, although it is often thought to be so.

An American patent is awarded to the individual person or persons who claim to be the inventor; in European countries a patent is often granted to a company, on the understanding that many persons within the company may have contributed to the invention.

The patent files contain thousands of lens designs, many of which have never been put into production. The designs provide useful suggestions as

VII. References

to how a particular problem may be solved. In this book a few patent numbers have been given as footnotes when the patent appears to cover a particular design fairly closely, but examination of actual lenses often reveals differences in detail from the example given in the patent. In these footnote references, a patent in the country of origin comes first, followed by duplicate patents in other countries in order: United States, Britain, and Germany. Patents in countries other than these are not listed, as I have little information about them. The stated date of a patent is the convention date, that is, the date of the earliest application in any country. Note that British patents filed prior to 1916 were numbered from 1 in each year, but after January 1916 a consecutive system of numbers was used, as in other countries.

Attempts to correlate lens structure with patent data often reveal that, as the years pass, a manufacturer may change the structure of a lens without changing its name and without any indication that a change has occurred. Sometimes the same name is revived much later to apply to an entirely different lens structure; examples are names such as Biogon, Ektar, and Planar, and there are many other instances of this having occurred.

VII. REFERENCES

A. Books Dealing with Photographic Lenses

D. van Monckhoven, *Photographic Optics*. Hardwicke, London, 1867.

E. Wallon, *Traité Elémentaire de L'Objectif Photographique*. Gauthier Villars, Paris, 1891.

J. Traill Taylor, *The Optics of Photography and Photographic Lenses*, 2d edition. Macmillan, New York, 1898.

M. von Rohr, *Theorie und Geschichte des photographischen Objectivs*. Springer, Berlin, 1899.

O. Lummer, *Contributions to Photographic Optics*, trans. S. P. Thompson. Macmillan, London, 1900.

E. Wallon, *Choix et Usage des Objectifs Photographiques*. Gauthier Villars, Paris, 1902.

C. Beck and H. Andrews, *Photographic Lenses*. Lund Humphries, London, ca 1903.

J. M. Eder, *Die photographischen Objektive*. Knapp, Halle, 1911.

W. Merté, R. Richter, and M. von Rohr, *Das photographische Objektiv*. Springer, Wien, 1932. Also Edwards, Ann Arbor, 1944.

CHAPTER 2

Meniscus Landscape Lenses

I. THE WOLLASTON SIMPLE MENISCUS

It is indeed an interesting paradox that the first photographic lens was designed some 25 years before the invention of photography. Before that time, practically all lenses were intended to cover a small angular field, such as the objective lenses of telescopes and microscopes. With the coming of the camera obscura, however, lenses were required to cover a wide flat field, but this problem had baffled the opticians of the eighteenth century.

About 1812 the English scientist W. H. Wollaston (1766–1828) discovered that a meniscus-shaped lens with its concave side to the front was capable of giving a much flatter field than the simple biconvex lens generally used in the camera obscuras of that time.[1] Two recent photographs illustrate his discovery (Fig. 2.1). The first was taken with a simple biconvex lens of 6 inches focal length and ¾ inch aperture; the poor quality of the outer parts of the image is evident. The second picture was made with a

---

[1] W. H. Wollaston, "On a periscopic camera obscura and microscope." *Phil. Mag.* **41**, 124 (1813).

(a)

(b)

Figure 2.1. Photographs taken with simple lenses: (a) a biconvex lens and (b) a meniscus lens.

meniscus-shaped lens of the same focal length, with a ¾ inch stop located a short distance ahead of the lens on the concave side. It is immediately obvious how great an improvement has been made by this small change in the shape of the lens. Wollaston's arrangement was used in the large camera obscura lenses mounted in the roof of a cabin and in the smaller tentlike devices popular with artists, as shown in Figure 2.2. The plane mirror used to bend the light downward also served to rectify the left-handedness of the image.

With the introduction of photography in 1839, it was found that Wollaston's simple lens suffered from excessive chromatic aberration. The image in blue light to which the daguerreotype plate was sensitive fell closer to the lens than the image in yellow light to which the eye was sensitive, so that with such a simple lens it was found to be impossible to focus the image sharply both on the ground glass and on the sensitive plate. It might be argued that if the magnitude of the chromatic focus difference were known, an appropriate separation between the ground glass and the plate holder could be introduced into the camera itself, but even this would not work, as the chromatic separation varies with the distance of the object from the camera. Nevertheless, the simple Wollaston lens was revived successfully in the late 1890s for use on a box camera in which there was no means for focusing the image visually on ground glass.

Actually, the lack of correction of both the chromatic and spherical aberration has the effect of increasing the depth of field of a simple box camera. The image of an object located at some particular distance from the lens is in focus for certain combinations of lens zone and wavelength of the light, while it is more or less out of focus for other zones and wavelengths. The picture formed on the film, therefore, consists of a sharp

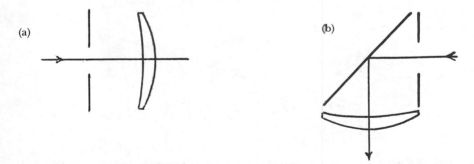

Figure 2.2. The Wollaston Landscape lens: (a) for a camera and (b) as used on a camera obscura.

image superposed on a slightly blurred image, but as the sharp image is likely to be brighter than the blurred image, it will be more strongly imaged on the film, particularly if the exposure is on the lean side. Over-exposure will, of course, record everything, resulting in a somewhat blurred and "mushy" negative.

II. THE ACHROMATIC LANDSCAPE LENS

While Daguerre was working on the development of his system of photography in the 1830s, he needed a suitable lens for his camera, so he turned to his friend Charles Chevalier for help. Chevalier was a well-known maker of telescopes and microscopes, and he had many lenses on his shelves that he could offer to Daguerre. The object glass of a telescope when used the right way round forms a sharp image lying on a curved surface. However, Chevalier noticed that if the lens is turned around with the flat side toward the front, the image becomes hazy but the field be-

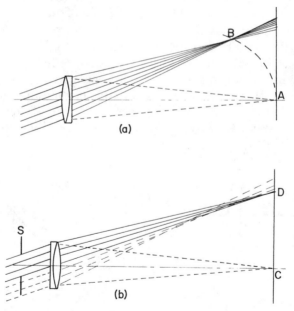

Figure 2.3. Path of rays through a telescope objective, entering at 20°; (a) correct way around and (b) reversed.

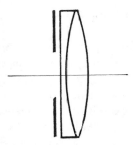

Figure 2.4. Daguerreotype lens by Lerebours and Secrétan.

comes flat, even out to some 20° from the axis. The true-to-scale images in Figure 2.3 illustrate these effects. In Figure 2.3(b), the dashed rays passing through the lower half of the lens tend to be bent sharply upward, giving a one-sided blur called coma, but these rays can be cut off by means of a stop placed a short distance ahead of the lens, as shown.

This was the lens that Chevalier manufactured for use on the official daguerreotype cameras sold by Alphonse Giroux in 1839. The focal length was 15 inches, the lens diameter 3½ inches, and stop opening 1 inch, the stop being mounted about 3 inches in front of the lens. The plate size was 16 × 22 cm, which became 6½ × 8½ inches in England where it is known as the whole-plate size to this day. Chevalier's lens was achromatized for visible light, which was a great improvement over the unachromatized Wollaston lens, although careful workers discovered that it would be better if the achromatism were centered about the blue actinic spectral region rather than in the visible. However, at the $f/15$ aperture used by Chevalier the error in achromatism was not important. It was found to be

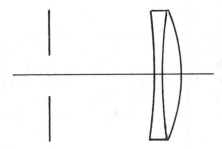

Figure 2.5. The French Landscape lens.

much more significant in the high-aperture portrait lenses introduced by Petzval and others.

At about the same time, the French optician N. M. P. Lerebours (1807 – 1873) made some lenses for the daguerreotype cameras manufactured by Gaudin. These lenses were back-to-front telescope objectives similar to those of Chevalier, except that a wheel of stops was placed close to the lens (Fig. 2.4). This, of course, greatly increased the field of view, but the stop no longer served to cut off the steeply sloping lower rays, and signs of coma appeared in the image. Lerebours also made cameras equipped with this lens, even after he became associated with Secrétan in 1845.

Chevalier quickly realized that if he made his landscape lenses somewhat meniscus in shape, following the suggestion of Wollaston, the stop could be moved closer to the lens and the angular field accordingly increased. The lens so constructed became known as the French Landscape lens (Fig. 2.5), and it was made by every manufacturer in a variety of mountings for nearly a century. Unfortunately, the strong cemented interface in these lenses had the effect of bending the field too far backward, so that it is questionable whether this achromatic lens or the simple Wollaston objective would give the better image. The achromatism was, of course, an advantage, but besides that there is little to choose between them. The Kodak Company fitted achromatic lenses to their better cameras for many years, and they also made some semiachromats in which the cemented interface was plane, but they finally abandoned achromatic landscape lenses after they found that the average user could not detect the difference.

III. OTHER MODIFICATIONS OF THE LANDSCAPE LENS

A. The Grubb Aplanat

A few attempts have been made over the years to improve the French Landscape lens, but without much success. In 1857 Thomas Grubb (1800 – 1878) made a lens that he called the Aplanat because of its low spherical aberration.[2] In this lens the meniscus crown element was placed in front of the meniscus flint element (Fig. 2.6). Because of the absence of spherical aberration the coma could not be reduced by a choice of stop

---

[2] Brit. Pat. 2,574/57.

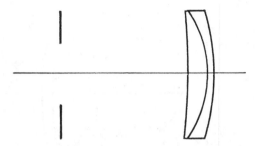

Figure 2.6. The Grubb Aplanat lens.

position, although the field could be readily flattened by this means. In spite of the inherent coma, many hundreds of these Grubb lenses were sold during the next forty years, and a similar type of construction has been used in some recent soft-focus portrait lenses.

Grubb's lens was so similar to the rear half of the Rapid Rectilinear lens announced in 1866 that one wonders whether Dallmeyer merely assembled two of Grubb's objectives about a central stop to make his famous lens (see Chapter 4, Section IVB).

B. Dallmeyer's Rapid Landscape Lens

In 1864 J. H. Dallmeyer (1830 – 1883) developed the Rapid Landscape lens consisting of three elements cemented together instead of the usual two (Fig. 2.7). The inventor claimed an aperture of $f/11$ covering a field of ±30° with this arrangement.[3] The type was subsequently used by other manufacturers, so it must have had some virtues. In 1880 Dallmeyer

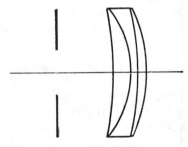

Figure 2.7. Dallmeyer's Rapid Landscape lens.

---

[3] Brit. Pat. 2,539/64; U.S. Pat. 61,812.

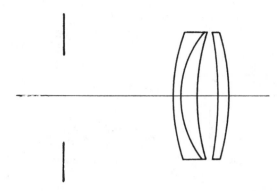

Figure 2.8. Dallmeyer's Rectilinear Landscape lens.

modified the lens to form the Rapid Landscape Lens, Long Focus, which covered a somewhat narrower field in the longer focal lengths. In the 1890s Ross issued a Triple Landscape lens that was similar in construction to the rear half of Gundlach's Rectigraphic lens (see Fig. 4.14).

C. Distortionless Landscape Lenses

It should be remarked that no matter how well a landscape lens is designed and made, its asymmetric construction inevitably leads to a noticeable amount of distortion, which shows up in pictures of architectural subjects. A man named J. T. Goddard[4] (died 1864) suggested in 1859 that the distortion could be corrected by inserting a zero-power cemented doublet between a landscape lens and the stop, but he had no manufacturing facilities, and it was left to T. R. Dallmeyer (1859–1906) to commercialize the design.[5] In 1888 Dallmeyer marketed the Rectilinear Landscape lens (Fig. 2.8) based on this principle. However, by that time many much better lenses were available, and the new lens did not last very long.

D. The Front Meniscus Lens

An important improvement in single lenses was made about 1934 when Kodak turned the simple Wollaston meniscus around and mounted the lens ahead of the stop. This greatly shortened the camera, which was its

[4] A biography of J. T. Goddard is given in M. von Rohr, *Theorie und Geschichte des photographischen Objektivs*, page 181. Springer, Berlin, 1899.

[5] Brit. Pat. 1,583/88.

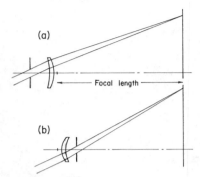

Figure 2.9. Comparison of Rear and Front Landscape lenses of the same focal length.

main purpose (Fig. 2.9), but it also served to protect the delicate shutter mechanism from dirt, and the user was likely to be impressed with the large bull's-eye in front of the camera. The use of a front landscape lens of this kind reverses the sign of the distortion, making it pincushion instead of barrel. Unfortunately, if the lens surfaces are reasonably weak, the field is found to be decidedly inward-curving, and in some cameras an attempt was made to compensate for this by the use of a curved film platen, but a cylindrical film is a poor fit on a spherically curved image field. More recently, the lenses on low-cost cameras have been made of plastic material rather than glass, and then, of course, the surfaces can be made much stronger, giving a field as flat as that of an ordinary rear landscape lens. However, the spherical aberration becomes worse with the stronger surfaces, but the overall advantages of a front meniscus lens are so great that this has now become the universal arrangement.

CHAPTER 3

Portrait Lenses

In 1839, when the daguerreotype process was announced, the only available lens was the achromat with front stop suggested by Charles Chevalier. The $f/15$ aperture of this lens was far too low for portraiture, as at that time an exposure of 30 minutes was required in bright sunlight.

Since photographers were naturally anxious to make portraits, something had to be done to speed up the process. The attack was made in two directions. The chemistry of the process was investigated, and by the addition of bromine the plates were rendered much more sensitive to light. At the same time opticians began to seek ways by which the lens aperture could be drastically increased. To be sure, telescope objectives of that era could be made with an aperture of $f/5$, but they covered only a very narrow field, and nobody knew how to widen the field without introducing impossibly large aberrations.

I. CHEVALIER'S PHOTOGRAPHE

Chevalier, always the empiricist, began experimenting with various combinations of lenses that he had available on the shelf and discovered that by adding another achromat between the stop and the lens of his

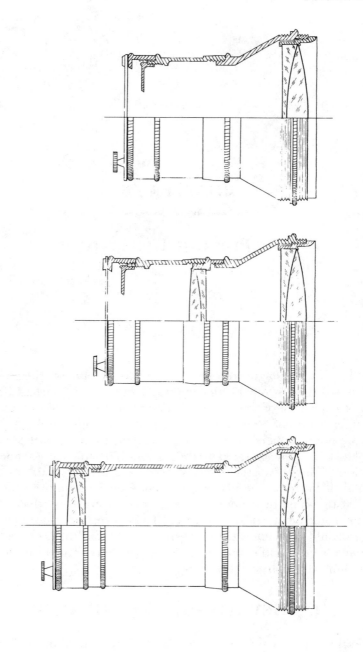

Figure 3.1. Three forms of Chevalier's Photographe à Verres Combinés.

landscape objective, he could raise the aperture to $f/6$, about six times faster than his original system.[1] Furthermore, he found that he could use different added lenses and even turn the system around in the camera to adapt it for various uses, landscapes, portraits, and so on (Fig. 3.1). He called it the Photographe à Verres Combinés à Foyer Variable, the word photographe referring to the lens and not to the picture made with it.

Unfortunately, Chevalier's new lens was not all that good, and although it was manufactured for some twenty years by Chevalier and his son, it could in no way compete with the Petzval Portrait lens. Very few remain today.

II. THE PETZVAL PORTRAIT LENS

The story goes that news of the daguerreotype process and of a prize being offered by the Society for the Encouragement of National Industry for a faster lens was conveyed to Vienna by A. F. von Ettingshausen (1796–1878), professor of physics at the university. He undertook to persuade his colleague Joseph Max Petzval (1807–1891), professor of higher mathematics, to attempt the task of designing such a lens.[2] Remarkably, since neither he nor anyone else knew anything about lens design, the 32-year-old Petzval accepted the challenge to design a flat-field lens fast enough for portraiture. For help with his calculations he approached the Archduke Ludwig, Director General of Artillery in the Austrian army, who ordered that "Corporals Löschner and Haim and eight gunners skilled in computing be placed at his disposal." With this somewhat unusual help, in only about six months Petzval completed the design of two objectives, both of which embodied the same conventional $f/5$ cemented doublet telescope objective in front with a separated achromat behind (Fig. 3.2). The two designs were a portrait lens of $f/3.6$ aperture and a wide-angle objective of $f/8.7$ aperture. (The latter was really a telephoto, as the rear component was of negative power; see Chapter 9, I.) Petzval sent both designs to his friend the optician P. W. F. Voigtländer (1812–1878) in Vienna, who made the portrait lens at once but kept the other design in his desk. The new Portrait lens was just what was needed; it was

---

[1] C. Chevalier, *Nouvelles Instructions sur l'Usage du Daguerreotype*, page 18. Chez l'Auteur, Paris, 1841.

[2] The story of the development of the Petzval lens is given in detail in J. M. Eder, *History of Photography*, trans. E. Epstean, pp. 291–313. Dover, New York, 1978.

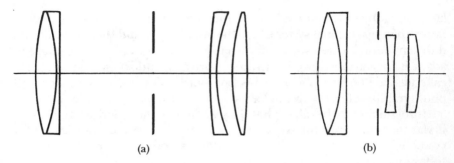

(a)   (b)

Figure 3.2. Two Petzval designs: (a) Portrait lens and (b) Orthostigmat.

about twenty times as fast as the Chevalier landscape lens used at that time on daguerreotype cameras.

In designing his portrait lens, it is probable that Petzval started out with a symmetrical arrangement of two identical telescope objectives spaced apart back to back, but he soon found that he had to separate the two elements in the rear doublet and bend them independently to correct the spherical aberration and coma. The power of the rear component was chosen to introduce just enough overcorrected astigmatism to flatten the tangential field.[3] As a consequence, the definition was excellent over a circular area in the middle of the picture, but it deteriorated gradually toward the outer part of the picture, as can be seen in many of the classical early portraits such as those by Julia Margaret Cameron. However, it can be argued that this was actually a good thing, as it tended to stress the portrait and suppress the unwanted background.

The first sample of the Portrait lens was completed in May 1840 and entrusted to A. F. C. Martin (1812–1882) of the physics faculty at Vienna for testing. The lens was found to be excellent, and it was submitted to the Society of Encouragement early in 1841 in competition with Chevalier's Photographe à Verres Combinés. The committee studied the two lenses at great length and finally in March 1842 awarded a platinum medal to Chevalier, mainly because of the convertible features of his lens, and a silver medal to Petzval. But time was on the side of Petzval, and by 1850 some 8,000 Petzval lenses had been made while the Chevalier lens was all but forgotten.

The original $f/3.6$ Petzval Portrait lens in a focal length of 150 mm was

---

[3] The construction of the original lens is given in M. von Rohr, *Theorie und Geschichte des photographischen Objektive*, p. 250. Springer, Berlin, 1899.

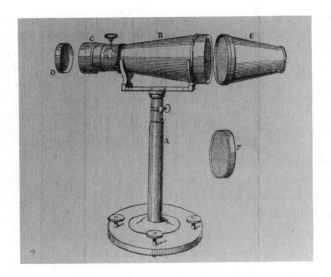

Figure 3.3. Original Voigtländer cylindrical metal camera.

fitted by Voigtländer into a conical metal camera equipped with a ground glass focusing screen and a viewing magnifier at the rear (Fig. 3.3). After focusing the camera on the subject, it was taken into the dark room, where the focusing screen was replaced by a circular daguerreotype plate and the camera was restored to its stand for the actual exposure. Some 70 of these cameras were made in 1841 and 600 in 1842, but today only three complete outfits are known to exist, in Munich, Vienna, and at the Voigtländer factory in Brunswick.[4] Original circular daguerreotypes made with this camera are known but exceedingly rare.

Petzval and Voigtländer soon quarreled, because Petzval complained that he was not being adequately recompensed for his invention. Indeed, within five years the two men ceased to be on speaking terms, and in 1854 Petzval turned to another Viennese optician, C. Dietzler, for help. Dietzler made a large number of Portrait lenses, but he could not prevent Voigtländer from making them also, as Petzval had only an Austrian patent and Voigtländer had moved to Brunswick in 1849 where Petzval had no control over his actions. Petzval's Portrait lens was copied freely by every optician under the name German System. These were generally mounted

---

[4] A color photograph of the outfit in its case is given in J. Willsberger, *The History of Photography*, p. 19. Doubleday, New York, 1977.

in polished brass tubes, but they usually bore no information as to the maker's name, the focal length, or the relative aperture. They were fitted to numerous cameras, both single and stereoscopic, commonly with a rack-and-pinion arrangement for focusing. They were also used extensively for slide projectors (Magic Lanterns). Petzval-type lenses were employed universally by portrait photographers up to the 1920s, by which time anastigmats with comparable aperture had become available.

The principal early manufacturers of Petzval lenses were:

England: Ross, Dallmeyer

France: Hermagis, Auzoux, Gasc & Charconnet, Derogy, Lerebours & Secrétan, Jamin & Darlot

Germany: Voigtländer, Steinheil, Busch

Switzerland: Suter

USA: Harrison, Holmes Booth and Haydens

A. The Cône Centralisateur Lens

An interesting example of a Petzval-type lens was that made by Jamin & Darlot in 1855 called the Cône Centralisateur. Jean Theodor Jamin commenced the manufacture of lenses and optical instruments in Paris in 1822. In 1855 he was joined by Alphonse Darlot (1828–1895), who had worked previously with Lerebours and Secrétan. In 1860 Jamin retired and handed over control of the company to Darlot; Jamin died in 1867. At that time the establishment was located at 14 rue Chapon, off the rue St. Martin in Paris. Darlot continued very actively making lenses and other optical devices until his death in 1895, when the factory was acquired by L. Turillon.

The Cône Centralisateur objective (Fig. 3.4) was a normal Petzval lens in a plain brass mount with a rack-and-pinion focusing mechanism in front, while behind the mounting flange the tube widened out into a black-painted cone supporting the rear component. The purpose of the cone was to prevent internal reflection of sunlight from falling on the sensitized plate. In some models the separation between the lenses could be altered by the user. A committee of the French Photographic Society examined the lens and found it to be satisfactory.[5]

---

[5] *Bull. Soc. Franc. de Phot.* **1**, 341 (1855).

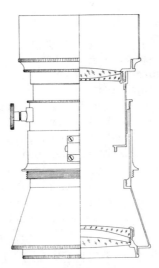

Figure 3.4. The Cône Centralisateur lens mount.

B. The Grün Liquid Lens

In 1901 Dr. Edward F. Grün of Brighton, England, patented a lens of the Petzval type in which the two rear elements were separated by a layer of oil in place of air. The good doctor, knowing nothing of conventional lens design, was convinced that oil had some magical ability to improve any lens; this idea probably derived from his experience with the oil-immersion microscope objective. He made many experiments with the introduction of oil into the airspaces of ordinary lenses with, of course, no useful results. The lens he actually offered for sale was a 7½-inch Petzval with an aperture of $f/3$ and oil between the two rear elements.[6] It did not perform as well as a conventional Petzval lens of the same focal length and aperture. So, we may ask, why use oil at all? Dr. Grün obtained a great deal of publicity and some adverse comments by publishing photographs of stage shows taken by ordinary theater lighting using one of his liquid lenses. By 1903 all interest in his lenses had disappeared.

---

[6] Brit. Pat. 6,194/01; U.S. Pat. 695,606. See also *Image*, George Eastman House **20,** 63 (September 1977).

C. The Reversing Prism

This might be a good place to mention the Reversing Prism that was sometimes mounted in front of the lens used on a daguerreotype camera, because without it the image would be reversed from left to right, and any lettering in the scene would appear in looking-glass language. Many existing daguerreotype images exhibit this type of reversal because most daguerreotype cameras were not equipped with a reversing prism or mirror. If this image reversal caused real trouble, the photographer could copy the picture on to another daguerreotype plate to hand to the customer. This problem, of course, disappeared when negative-positive processes became universal. In modern instant cameras such as those constructed by Polaroid and Kodak, it is necessary to incorporate one or two mirrors into the camera to effect the necessary left-right reversal.

III. OTHER PORTRAIT LENSES

A. The Dallmeyer Patent Portrait Lens

In the design of a portrait lens of the Petzval type it is immaterial whether the crown or the flint element in the rear component comes first. Petzval chose to have the flint in front, as shown in Figure 3.2(a). In 1866 J. H. Dallmeyer patented a lens of the other type in which the crown element came in front of the flint (Fig. 3.5).[7] There is actually little to choose between the two types, some manufacturers preferring one type and some the other. Indeed, some companies made both types. For instance, the Bausch & Lomb $f/2.2$ Series B objectives were of the Petzval type, while their $f/4$ Series A lenses were of the Dallmeyer type. In his patent Dallmeyer suggested that a soft-focus effect could be introduced into the

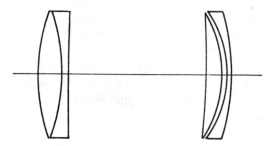

Figure 3.5. Dallmeyer's Patent Portrait lens.

---

[7] Brit. Pat. 2,502/66; U.S. Pat. 65,729. See also *Brit. J. Phot.* **13**, 606 (1866).

Figure 3.6. The Zincke-Sommer lens.

image by varying the separation between the two rear elements, the Bausch & Lomb Series A lenses being provided with a screw adjustment for this purpose.

In 1870 the aperture of the Dallmeyer type was raised to $f/2.37$ by H. F. A. Zincke-Sommer,[8] half-brother of F. R. von Voigtländer (Fig. 3.6). In 1858 Dallmeyer himself had achieved a similar aperture in the little lens used by Thomas Skaife in his Pistolgraph camera.

B. The Voigtländer Lens of 1878

In 1878 F. R. von Voigtländer tackled anew the design philosophy of the Petzval Portrait lens. He argued that there is really no need for the front component to be fully corrected for spherical aberration and that, if he were to bend the front component suitably, it should be possible to create an objective in which both components could be cemented.[9] This he succeeded in doing, starting with the Dallmeyer type (Fig. 3.7).

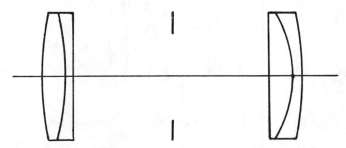

Figure 3.7. Voigtländer's portrait lens of 1878.

---

[9] Ger. Pat. 5,761; Brit. Pat. 4,756/78.
[8] Von Rohr, ref. 3, page 275.

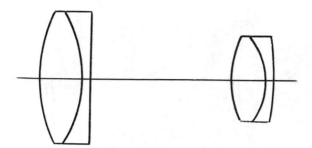

Figure 3.8. A typical $f/1.6$ movie projection lens.

It is not known how many of these objectives were sold. However, most of the lenses that have been used on 8 mm and 16 mm movie projectors were of this general type, though considerably modified from Voigtländer's original design. In Figure 3.8 is shown an $f/1.6$ objective covering ±7° that was patented by W. F. Repp in 1922.[10]

C. The Steinheil Portrait Antiplanet

In 1881 Dr. H. A. Steinheil (1832–1893) designed a modified Petzval lens in which the strong rear flint element was widely spaced from the rear crown (Fig. 3.9), giving a somewhat negative focal length to the rear component and therefore requiring an unusually strong front component.[11] The opposition of the positive and negative powers led to the name

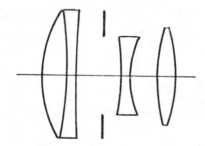

Figure 3.9. The Steinheil Portrait Antiplanet.

---

[10] U.S. Pat. 1,479,251 (1922).

[11] U.S. Pat. 241,438; Brit. Pat. 1,602/81. See also J. M. Eder, *Die photographischen Objektive*, p. 81. Knapp, Halle, 1911.

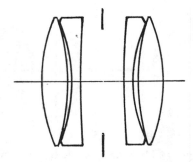

Figure 3.10. The Kodak $f/1.9$ cine lens.

Antiplanet. The spherical aberration was well corrected at $f/3.8$ and the field was somewhat inward-curving and almost free from astigmatism at 15° from the axis. Few, if any, examples of this lens exist today.

D. The Kodak $f/1.9$ Cine Lens

One of the most successful modifications of the Petzval type was that designed by C. W. Frederick (1870–1942) and F. E. Altman (1893–1964) in 1923 (Fig. 3.10), primarily for use on the Kodak 16 mm movie camera.[12] The front and rear lens pairs were in edge contact, and the refractive index of the elements was the same, namely, about 1.62. The lens covered ±14° at $f/1.9$, and many thousands were manufactured in the 1-inch size. A few 2-inch lenses were also made as projection lenses, particularly for the lenticular Kodacolor process of 1928 where the low vignetting was a great help.

E. The R-Biotar

In 1932 Willy Merté (1889–1948) of Zeiss designed a very high-aperture narrow-angle lens, primarily for making 16 mm movies from the image on an X ray fluorescent screen.[13] This was the $f/0.85$ R-Biotar (R for Roentgen) made in 45 mm and 55 mm sizes. It was basically a modified Petzval lens, as can be seen from the diagram in Figure 3.11.

---

[12] U.S. Pat. 1,620,337 (1923).
[13] Ger. Pat. 607,631; U.S. Pat. 1,967,836 (1932).

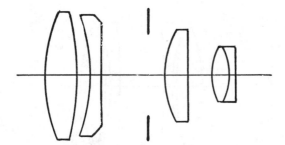

Figure 3.11. The Zeiss R-biotar lens.

IV. CURVED-FIELD LENSES

It was pointed out in Chapter 1 that a positive lens naturally has an inward-curving field, which in the absence of all other aberrations falls into the Petzval surface. Most of a lens designer's efforts are, therefore, directed toward trying to overcome this defect and move the image back into a flat plane without introducing too much astigmatism in the process. However, if a curved field can be used, then it is not difficult to design a lens in which all the important aberrations are well corrected at a relatively high aperture. Such a lens was designed and built[14] by R. E. Hopkins and D. P. Feder in 1948 at the University of Rochester (Fig. 3.12). The lens had a focal length of 6 inches and an aperture of $f/1$, so that the aperture was also 6 inches in diameter. In the camera the film was forced against a curved platen over a field of $\pm 20°$ by gas pressure.

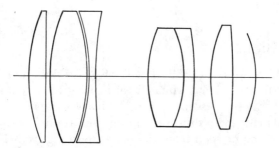

Figure 3.12. The Hopkins and Feder $f/1$ objective.

---

[14] *Illus. London News* **213**, 626 (December 1948).

Other well-known types of curved-field lenses are the Sutton ball lens described in Chapter 4, Section II, and the Schmidt camera (Chapter 12, Section IIA).

V. THE FIELD FLATTENER

In 1873 C. Piazzi Smyth (1819–1900), the Astronomer Royal for Scotland, discovered a way to remove the astigmatism and field curvature of an ordinary Petzval Portrait lens, defects that had seriously bothered him while taking photographs in Egypt seven years earlier. He argued that since a plate of glass inserted between a lens and its image has the effect of moving the image away from the lens by an amount equal to about one-third of the thickness of the plate, by inserting a plano-concave lens at the image plane the greater thickness at the edge as compared to the center would have the effect of flattening the field.[15] Unfortunately, the maker of his Petzval lens had already tried to flatten the field by deliberately introducing some overcorrected astigmatism, but Smyth soon found that reducing the central airspace in the lens would eliminate this astigmatism, leaving his field-flattener lens to flatten the field without introducing any astigmatism, even over a field of $\pm 17°$ at $f/3.6$. It is, of course, preferable to design the whole system as a unit, rather than merely add a field flattener to a well-designed objective.

The virtues of the field flattener were ignored for many years. Von Rohr designed an $f/1.9$ lens with a field flattener in 1911 (Fig. 3.13);[16] and

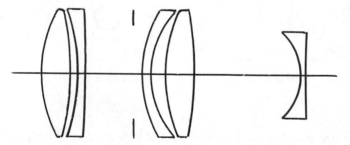

Figure 3.13. Von Rohr lens with a field flattener.

---

[15] *Brit. J. Phot.* **22**, 208 (1875). See also *Brit. J. Phot. Almanac,* 1874, p. 43.
[16] *Zeits. für Instkde.* **31**, 265 (1911).

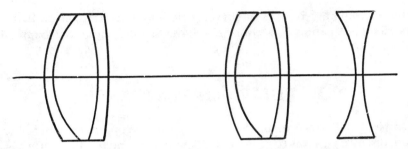

Figure 3.14. H. D. Taylor's $f/2$ lens with field flattener.

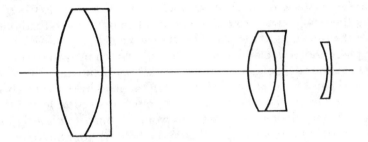

Figure 3.15. Kodak cine projection lens with field flattener.

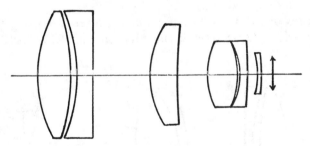

Figure 3.16. Kodak Projection Ektar $f/1$ lens.

in 1917 Dennis Taylor designed an $f/2$ objective consisting of two cemented triplets followed by a field flattener (Fig. 3.14).[17] This objective
was constructed by Taylor-Hobson and used for stellar photography at
Mount Wilson Observatory.

In 1934 a field flattener was added to the $f/1.6$ lenses used for the projection of small motion pictures (Fig. 3.15).[18] More recently the aperture has been raised to $f/1$ by introducing a positive meniscus element between the two doublets, as well as a field flattener in the rear (Fig. 3.16).

A few lenses for aerial cameras have incorporated a field flattener, often polished on the anterior surface of the glass platen used to keep the film flat in the camera.

[17] Brit. Pat. 127,058 (1917). See also *Trans. Opt. Soc. (London)* **24**, 148 (1923).
[18] U.S. Pat. 2,076,190 (1934). See also *JSMPTE* **54**, 337 (1950).

CHAPTER 4

Early Double Objectives

I. THE ADVANTAGES OF SYMMETRY

It was discovered very early that symmetry about a central stop confers many benefits on the lens designer. The first of these was found almost immediately after the introduction of photography, namely, that distortion is automatically corrected if a symmetrical construction is employed. Simple observation showed that a meniscus lens with stop in front exhibits barrel distortion, whereas if it is turned around with the stop behind the lens, the distortion reverses sign and becomes pincushion. Hence if a pair of similar lenses was mounted one in front of and one behind a central stop, the two distortions would cancel each other and leave a distortionless system.

Thomas Davidson, in 1841, assembled two Chevalier achromatic menisci in this way, producing his Combination Lens or Symmetrical Doublet, and in 1844 G. S. Cundell similarly mounted two simple Wollaston lenses about a central stop. Many early cameras were equipped with a pair of simple or achromatic menisci put together in this manner without any attempt at a proper design.

We may wonder why distortion was regarded as such an important aberration at a time when photographers were barely able to make pictures at all. This must have been because the extreme slowness of the original daguerreotype process drove photographers to choose stationary objects for their early efforts. What subject could be more suitable than a fine building? The straight lines of walls and roof would appear noticeably curved if a distorting lens were used. Landscapes and portraits, of course, would not show the effects of distortion, but architectural subjects most certainly would. The theory of distortion was fully worked out by R. H. Bow and T. Sutton in 1860.[1]

It can readily be shown that, besides distortion, two other aberrations are automatically corrected by a symmetrical construction, namely, lateral color (chromatic difference of magnification) and coma.[2] The lens designer could therefore ignore these difficult aberrations when designing the half-system of a symmetrical objective, confident that symmetry would remove them. It must not be forgotten that symmetry also implies equality of the conjugate distances, so that a symmetrical system must be used at unit magnification if the three transverse aberrations are to be completely removed. Actually, most symmetrical objectives exhibit only very small transverse aberration residuals even when used with an infinitely distant object, so the convenience of the symmetrical form remains valid. On the other hand, it has been found that unsymmetrical designs can be made to have much greater apertures and cover wider fields than symmetrical forms, and today the only perfectly symmetrical objectives in regular use are those intended to be used at or close to unit magnification, such as the large process lenses used in the graphic arts and for map copying, where distortion cannot be tolerated. The lenses used in Xerox-type copying machines are usually symmetrical. Some of the lenses used for printing motion pictures are of a symmetrical type of construction. Most enlarging lenses are nearly symmetrical, being adjusted to give optimum definition at a magnification between 2× and 10×.

II. EARLY SYMMETRICAL LENSES (NOT SPHERICALLY CORRECTED)

A. The Sutton Panoramic Lens

The first really wide-angle lens to be offered for sale was the water-filled Panoramic ball lens (Fig. 4.1) designed by Thomas Sutton (1819–1875) in

---

[1] *Brit. J. Phot.* **8**, 417 (1861); *Phot. Notes* **3**, 250 (1858) and **7**, 3 (1862).
[2] R. Kingslake, *Lens Design Fundamentals*, p. 203. Academic Press, New York, 1978.

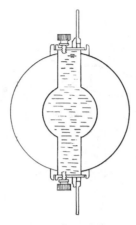

Figure 4.1. The Sutton Panoramic lens.

1859.[3] This was a monocentric design in which all the lens surfaces and the image surface were concentric about a single point in the middle of the lens. Such a system has no definite axis, and the axial and oblique images are identical and indistinguishable. No coma or astigmatism can possibly exist in such a lens, and the image height, measured along the curved plate, is proportional to the field angle in the object space. The angular field of Sutton's lens was wide, limited by vignetting to about ±60°, but the aperture was only $f/30$ because of the lack of correction for spherical and chromatic aberrations. The photographic plate for this lens should have been spherical with a radius of curvature equal to the focal length, but in practice a compromise was reached by the use of a cylindrical daguerreotype plate or by a pair of cylindrical glass plates fitting together with an emulsion-coated paper negative between them, which could be flattened out later for processing and printing. A few of these interesting lenses still exist, often manufactured by Ross, and they are valuable collector's items.

The idea of a ball lens was revived by J. G. Baker of Boston about 1942. Baker used a low-index glass core inside an outer shell of dense flint glass (Fig. 4.2). The system was monocentric but not symmetrical, the two inner radii of curvature being chosen to correct the chromatic and spherical aberrations, respectively. In this way excellent image quality was achieved at an aperture of $f/3.5$. The image, of course, lay on a spherical surface of radius equal to the focal length of the lens, and only a spherical photographic plate could be used. The shutter presented some awkward prob-

---

[3] Brit. Pat. 2,193/59. See also *Phot. J.* **6**, 184 (1860).

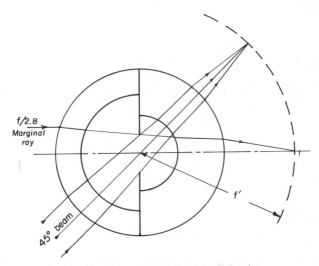

Figure 4.2. The Baker Ball lens.

lems, but these were overcome in an experimental camera with which some excellent aerial photographs were obtained. A resolving power of 200 lines per millimeter was claimed.[4]

The recently developed Zeiss Hologon lens of 1967 could be regarded as a modified ball lens, designed to be used on a flat film (see Chapter 10, Section III). It was made in a focal length of 15 mm at $f/8$ for use on a 24 mm × 36 mm format, representing an angular semifield of 55°.

B. The Globe Lens

The best known of the early wide-angle symmetrical lenses was undoubtedly the Globe lens, patented in 1860 by C. C. Harrison and J. Schnitzer of New York (Fig. 4.3(a)).[5] This lens got its name from the fact that the external front and rear surfaces formed parts of the same sphere, although this peculiar property had no optical significance. Indeed, it was soon found that an objectionable ghost image could be removed by closing the central airspace a little, without in any way affecting the definition. The Globe lens was manufactured in large numbers and in various sizes by Harrison and several European manufacturers. It covered a semifield of

---

[4] *Image,* George Eastman House **22,** 31 (Dec. 1979).

[5] U.S. Pat. 35,605; Brit. Pat. 2,496/60.

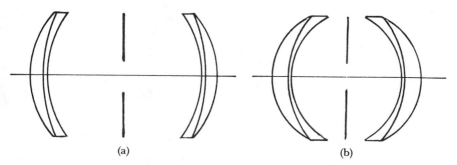

(a)   (b)

Figure 4.3. (a) The Harrison Globe lens and (b) the Busch Pantoskop.

almost 40° at the low aperture of $f/30$; the image was ultimately limited by a residue of astigmatism. Harrison died in 1864, but the Globe lens remained popular for many years. The widow of Henry Fitz, the American telescope maker, patented posthumously a modification of the globe lens in which the two cemented interfaces were made plane to simplify manufacture.

C. The Pantoskop

A similar but much better lens than the Globe was the Pantoskop, designed in 1865 by Emil Busch (1820–1888).[6] The components were more deeply curved than in the Globe (Fig. 4.3(b)), and the flat field was completely free from astigmatism out to ±40° at $f/25$. It was manufactured by Busch in seven sizes, from 52 mm to 540 mm in focal length; it was sold for many years as an excellent wide-angle lens. The skill exercised in the manufacture of the extremely thin and delicate meniscus elements constituting the lens was indeed remarkable, a task that would challenge the best optical craftsmen today.

D. The Periskop

Also in 1865, C. A. Steinheil (1801–1870) patented the Periskop, a properly designed $f/15$ combination of two single meniscus elements mounted about a central stop, which was similar to Cundell's arrangement (Fig. 4.4).[7] The lens shapes and the stop position were chosen to yield a flat

---

[6] M. von Rohr, *Theorie und Geschichte des photographischen Objektivs*, p. 281. Springer, Berlin, 1899.

[7] Brit. Pat. 2,937/65.

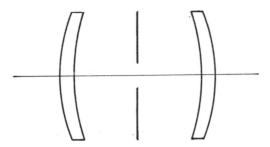

Figure 4.4. The Steinheil Periskop lens.

field, and symmetry took care of the coma and other transverse aberrations. The lens was not achromatic nor was it corrected for spherical aberration, and it did not succeed in replacing the Globe lens in popularity. Symmetrical lenses of this general type are called Periscopic to this day, and they have been frequently used on moderate-priced cameras.

In 1933 D. L. Wood of the Kodak Company patented a modification of the Periscopic type in which the front component was split into two parts by a plane-parallel airspace, the width of which could be varied by a screw to focus on close objects (Fig. 4.5).[8] This lens was known as the Bimat or Twindar, and it was fitted to several Kodak cameras.

E. The Hypergon

The limit of extreme wide-angle lenses covering a flat field was reached by the Goerz Hypergon lens of 1900.[9] This was a small lens of generally

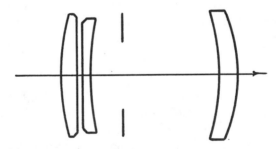

Figure 4.5. The Kodak Bimat and Twindar lenses.

---

[8] U.S. Pat. 1,954,340 (1933).
[9] Ger. Pat. 126,500; U.S. Pat. 706,650; Brit. Pat. 14,487/00.

Figure 4.6. The Goerz Hypergon lens.

ball-like exterior form, each component consisting of a deep meniscus element with almost equal radii on its two surfaces (Fig. 4.6). For example, in a lens of 100 mm focal length, the radii of curvature of the half-system were 8.510 mm and 8.471 mm, respectively, the thickness being 2.206 mm. The Petzval sum was practically zero, and the lens covered a flat anastigmatic field out to ±67°. The aperture was limited to $f/20$ because of the lack of correction of spherical and chromatic aberrations.

As a result of the drastic fall-off in illumination at the limits of this extremely wide field, it was found necessary to hold back the central exposure by means of a little cogwheel with long teeth (Fig. 4.7) mounted in front of the lens and spun by an air-bulb during exposure. After about five-sixths of the desired exposure time, the little cogwheel was swung aside to expose the middle of the field.

Figure 4.7. The Hypergon showing the spinning cogwheel partly open.

F. Wollensak Portrait Lenses

In the early years of the present century the Wollensak Optical Company in Rochester, New York, manufactured two soft-focus Portrait lenses, the $f/6$ Veritar and the $f/4$ Verito. These consisted of simple uncorrected lenses of the Periscopic type and yielded the soft-focus effect admired by portrait photographers at that time. The degree of softness could be varied by stopping down the lens. A similar soft-focus lens was made by Pinkham and Smith of Boston.

III. UNSYMMETRICAL DOUBLETS (NOT SPHERICALLY CORRECTED)

A. Early Experiments

For some reason, perhaps in an attempt to economize, several early workers tried departing from symmetry and designed lenses in which one-half was a cemented doublet while the other half was a single lens. For example, the English microscope maker F. H. Wenham, in 1857, proposed a lens in which the front component was a plano-convex single element while the rear was an overcorrected doublet, both lenses having their weak or plane sides facing the central stop. He regarded this as being a possible modification of the Petzval type. Another attempt was made by the American C. B. Boyle in 1865 in his Cornea Triplet. This consisted of a pair of meniscus landscape lenses, a singlet in front and a Chevalier-type cemented doublet behind.

B. The Ross Doublet

As early as 1841 Andrew Ross (1798–1859) had made an $f/4$ portrait lens for the London artist and calotypist Henry Collen (Fig. 4.8). This consisted of a combination of two widely spaced dissimiliar cemented doublets, an ordinary telescope objective in front and an unusual cemented doublet in the rear having a plano-convex crown element leading. As a portrait lens this was not a success, but later, in 1864, Ross's son Thomas (1818–1870) revived this type of construction to make a low-aperture distortionless lens (Fig. 4.9).[10] The Ross Doublet appeared in three variations:

---

[10] von Rohr, ref. 6, page 177.

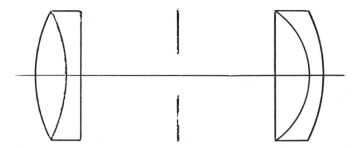

Figure 4.8. The Ross Collen lens.

Rapid, Actinic, or Instantaneous Doublet, covering ±28° at f/9
Medium-angle Doublet covering ±35° at f/15
Wide-angle Doublet covering ±40° at f/19

The construction of these lenses was similar, but as the field was reduced the lens became longer and less strongly meniscus in shape. These objectives were probably satisfactory, but they were soon replaced by the Rapid Rectilinears that appeared within a couple of years after the announcement of the Ross Doublet.

C. Hemisymmetrical Doublets

Many early workers felt that there was some advantage in constructing a double objective having components that were similar but scaled to different focal lengths, the weaker lens in front and the stronger lens behind. Such a system was called *hemisymmetrical*. One example was the Ratio Lens

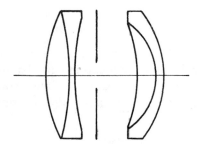

Figure 4.9. The Ross Doublet objective.

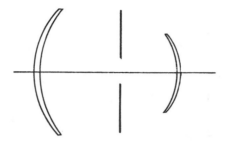

Figure 4.10. The Zentmayer lens.

patented in 1866 by C. B. Boyle of the Scovill Company in New York.[11] In this design the front component was a simple Wollaston meniscus, which was planned to be used with one of a series of interchangeable smaller but similar elements behind.

A somewhat similar arrangement was suggested in 1866 by the well-known microscope maker Joseph Zentmayer (1826 – 1888) of Philadelphia. In his patent he specified a pair of thin simple meniscus elements with a stop located at or close to the centers of curvature of the outer surfaces (Fig. 4.10).[12] In his catalog he listed a set of seven similar elements in interchangeable mountings, each lens having a focal length 1½ times the focal length of the next lens in the series. Thus, with lenses having focal lengths of 4, 6, 9, 13, 20, 30, and 45 inches he could form a set of hemi-symmetrical double objectives. By combining each lens in front with the next stronger lens in the rear, he could make combinations having the following approximate focal lengths: 2½, 3½, 5⅓, 8, 12, and 18 inches. This system was listed in Zentmayer's catalogs as late as 1899.

IV. SYMMETRICAL DOUBLETS (SPHERICALLY CORRECTED)

A. Achromatized Periscopic Type

This early type, which has no particular name, was often constructed empirically by assembling two identical cemented doublets about a central stop. The components might have been ordinary telescope objectives, as the aim of this arrangement was to eliminate distortion. In some cases the

[11] U.S. Pat. 52,129 (1866).
[12] U.S. Pat. 55,195 (1866).

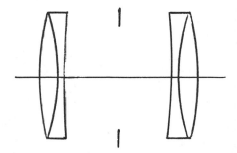

Figure 4.11. The Busch Nicola Perscheid objective.

aperture was as high as $f/4$, but on many early cameras made between 1880 and 1900 the aperture was more likely to be $f/8$ or $f/10$. This type was used in the soft-focus $f/4.5$ Nicola Perscheid lens manufactured by Busch in 1921 (Fig. 4.11).

B. The Rapid Rectilinear or Aplanat

One of the most important photographic objectives ever made was the famous Rapid Rectilinear or Aplanat of 1866. This design came midway between the invention of photography in 1840 and the introduction of the Anastigmat in 1890, and lenses of this type were fitted to all the better cameras for nearly sixty years, a record scarcely surpassed by any other lens.

By 1865 photographers had three types of lenses available to them: the simple landscape meniscus, the Petzval Portrait lens, and the wide-angle Globe lens or the Ross Doublet. What they needed was an intermediate lens covering about $\pm 24°$ at $f/6$ or $f/8$, which, of course, had to be free from distortion.

The Rapid Rectilinear lens was introduced by J. H. Dallmeyer in 1866. We do not know what led him to this highly successful design, but it may have been an assembly of two Grubb-type landscape aplanats about a central stop. Dallmeyer's patent showed a lens that was manufactured and sold under the name of Wide-angle Rectilinear.[13] The front and rear components were similar but not identical, the front being larger than the rear, as shown in Figure 4.12(a). Very soon Dallmeyer found that it was better to make the two halves identical (Fig. 4.12(b)), and this arrange-

---

[13] Brit. Pat. 2,502/66; U.S. Pat. 79,323.

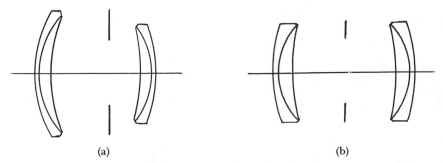

(a)   (b)

Figure 4.12. (a) The Dallmeyer Wide-angle Rectilinear lens and (b) the Rapid Rectilinear.

ment became the well-known Rapid Rectilinear. Most previous rectilinear (i.e., distortionless) lenses had been of low aperture, and Dallmeyer was therefore justified in calling his lens rapid, although the aperture was only $f/8$ or $f/6$ at the most.

Simultaneously and independently an almost identical design appeared in Germany called the Aplanat. This was designed and manufactured by Dr. H. A. Steinheil (1832 – 1893). As Steinheil and von Seidel (the mathematician who had recently established the theory of lens aberrations) were good friends, it is probable that the Aplanat had been designed on proper scientific principles, and Steinheil naturally supposed that Dallmeyer had pirated his invention. The argument became heated, and letters from both parties appeared in the scientific journals. When the smoke cleared it appeared that Steinheil had priority but by only a few weeks. Simultaneous inventions are actually quite common. The need is there, the necessary technology has been developed, and we must expect to find several inventors in various countries all working along similar lines.

The real clue to the construction of the Rapid Rectilinear lay in the choice of glass. The two glass types should differ as much as possible in refractive index yet be close in dispersive powers. The lower-index positive elements were inside, close to the stop, while the higher-index negative elements were outside. Among the glasses available in 1866 there were only flint glasses that met the requirements, and both Dallmeyer and Steinheil selected two flints, one light and one dense, to make their lenses. If the two glasses were widely different in properties the lens became long and worked at a higher aperture, while if the glasses were closely similar the lens became more compact, the components more meniscus-shaped, and the angular field became larger, though at a smaller aperture. Several series of these lenses differing in aperture and field therefore became

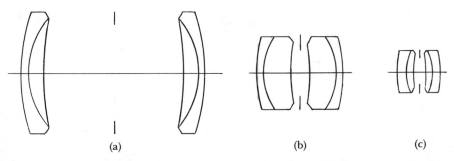

Figure 4.13. Three forms of Steinheil Aplanat, drawn to the same scale: (a) Rapid, $f/6.2$; (b) Landscape, $f/10$; and (c) Wide-angle, $f/18$.

possible.[14] Thus, Steinheil's Landscape Aplanat covered $\pm 30°$ at $f/10$; his Wide-angle Aplanat covered $\pm 45°$ at $f/18$; and he later developed a Portrait Aplanat, which covered $\pm 12°$ at $f/3.5$ (Fig. 4.13).

After the development of barium glasses, other combinations of glass types became possible that still met the requirements of a low V-difference and a high n-difference. For example, the Universal Aplanat of Steinheil used a common crown glass with a dense barium flint, and most of the later Rapid Rectilinear lenses also employed this type of glass combination.

The Rapid Rectilinear or Aplanat was a fabulously successful lens, and the basic design was immediately adopted by every manufacturer. The variety of trade names was bewildering, to say the least. Some of the trade names were:

Aplanascope	Leucograph	Polynar
Aplanatic	Lynkeioscope	Rectigraph
Aristoplanat	Monoplast	Rectilinear
Aristoscope	Orthoscope	Rectiplanat
Barytaplanat	Panoramic	Sphariscope
Biplanat	Pantoscope	Symmetrical
Eidoscope	Paraplanat	Universal
Euryscope	Perigraphic	Versar
Grossar	Planatograph	Voltas
Lamprodynast	Platystigmat	

---

[14] J. M. Eder, *Die photographischen Objektive,* pages 52–61. Knapp, Halle, 1911.

Figure 4.14. The Gundlach Rectigraphic.

Actually, the Rapid Rectilinear is a fairly expensive lens to manufacture. The internal cemented interface is strongly curved and has to be polished one on a block for both elements. The lens did have one outstanding virtue, in that the aberrations were well maintained over a wide range of object distances, so that the same lens could be used equally well on a camera or an enlarger, or for copying full-size without distortion.

C. The Gundlach Rectigraphic Lens

In 1890 Ernst Gundlach, then living in Rochester, New York, patented a variant of the Rapid Rectilinear in which he used three cemented elements in each half instead of two (Fig. 4.14).[15] The numerical construction of this lens was not disclosed, but probably its performance did not differ significantly from that of the Rapid Rectilinear.

V. UNSYMMETRICAL DOUBLETS (SPHERICALLY CORRECTED)

Presumably as an economy measure, several manufacturers have tried to combine a cemented doublet of the Rectilinear type with a single meniscus element, around a central stop. The absence of chromatic correction in the single element would lead to some degree of lateral color in the image, but this might not prove too serious. In 1894 H. L. H. Schroeder patented such a combination using the front of a Rapid Rectilinear and the rear of a

---

[15] U.S. Pat. 442,251; Brit. Pat. 12,741/90.

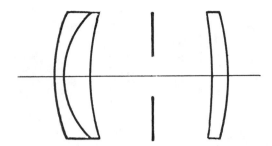

Figure 4.15. A typical Bausch & Lomb Semi-Rectilinear lens.

simple Periscopic.[16] This arrangement would work well if the front combination were sufficiently overcorrected to offset the residual aberrations of the uncorrected rear element. Bausch & Lomb supplied a number of lenses of this type to Kodak during the early 1900s. The lens was known commonly as a Semi-Rectilinear (Fig. 4.15).

A. Morrison's Lenses

This is a good place to mention a number of unusual designs manufactured by Richard Morrison of New York between 1872 and 1880. These lenses contained two separated components, some of which were singlets, some cemented doublets, and some airspaced. These lenses did not carry any names, but Morrison referred to them in his catalogs by such names as Wide-angle View Lens, Quick-working Triplet, Rapid Doublet, or Group Lens. However, these unusual types did not appear to be any better than more conventional designs.

VI. AIRSPACED TRIPLE COMBINATIONS

During the early years of photography several workers tried the effect of inserting a negative element between a pair of positive lenses. An example of this was the Triplet made by A. Ross in 1841 for Fox Talbot. A similar lens was proposed by G. Shadbolt in 1856. F. Scott Archer, in 1853, tried placing a negative element in the middle of a Petzval Portrait lens to increase the focal length. Thomas Sutton similarly made a Symmetrical

---

[16] U.S. Pat. 554,737 (1894).

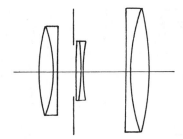

Figure 4.16. The Dallmeyer Triple Achromatic lens.

Triplet or Architectural View Lens in 1859 by mounting a negative element in the central stop of an achromatized Periscopic lens. Any of these efforts could have anticipated the famous Cooke Triplet, but in every case the components were so weak that they did not serve to reduce the Petzval sum.

A. Dallmeyer's Triple Achromatic Lens

This interesting objective was designed in 1861 by J. H. Dallmeyer.[17] It consisted of three rather weak achromats, in order positive negative positive, as shown in Figure 4.16. The lens worked at $f/10$, the definition was good, and the system was distortionless. It was manufactured in a range of focal lengths and appears to have been quite popular until it was replaced by the Rapid Rectilinear in 1866. A similar lens was made by T. Ross under the name of Actinic Triplet.

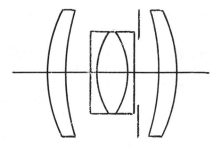

Figure 4.17. The Abbe Apochromatic Triplet.

---

[17] von Rohr, ref. 6, page 170.

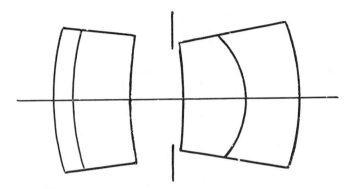

Figure 4.18. The Steinheil Group Aplanat.

B. The Abbe-Rudolph Apochromatic Triplet

In 1890 E. Abbe and P. Rudolph of the Zeiss Company turned their attention from microscope objectives to the design of a photographic lens.[18] They had been working on the correction of secondary spectrum in the microscope, and they felt that a similar procedure could be applied to photography. Their lens consisted of a thick cemented triplet inserted in the middle of a symmetrical Periscopic lens, as shown in Figure 4.17. Unfortunately, although the axial image was well corrected, the Petzval sum was much too large and the oblique images suffered from severe astigmatism. It is doubtful if any of these lenses were sold, but the effort did turn Rudolph's attention to the problems involved in his later design of the Anastigmat lens.

VII. THE STEINHEIL ANTIPLANETS

The Aplanat or Rapid Rectilinear lens of 1866 was symmetrical, each half having about twice the focal length of the double objective. As has been explained, symmetry confers the benefit of automatic correction of coma, distortion, and lateral color, but it does nothing else. Further, symmetry implies equality of the conjugate distances, and it is not surprising to find that a symmetrical lens when used with a distant object no longer

---

[18] Ger. Pat. 55,313; U.S. Pat. 435,271; Brit. Pat. 6,029/90.

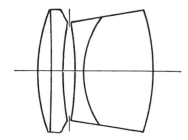

Figure 4.19. The Steinheil Group Antiplanet.

exhibited perfect correction for these three transverse aberrations. It therefore occurred to H. A. Steinheil that a somewhat unsymmetrical system might be actually better than a symmetrical arrangement, particularly in its correction for the remaining longitudinal aberrations and for its normal use with a very distant object.

His first experiment in this direction was the Group Aplanat of 1879, shown in Figure 4.18. The front component of this $f/6.2$ lens was somewhat weaker than the rear component, and both components were so thick that they had to be cone-shaped to avoid excessive vignetting at a field of $\pm 28°$. Judging by the published data given by von Rohr, the tangential field with a distant object was slightly backward-curving and the spherical aberration was slightly undercorrected, but the coma was less than in the symmetrical Aplanat.[19] The types of glass were the same as before.

However, Steinheil was still not satisfied, and he decided to go much further in the direction of asymmetry and actually reverse the roles of crown and flint glasses in the front component. Making the positive element of high-index flint glass and the negative component of low-index crown led to horrendous undercorrection of the spherical aberration and an inward-curving field. These defects were compensated by a strongly overcorrected rear component of low power, which served merely as an aberration corrector. In this way Steinheil developed the Group Antiplanet of 1881 (Fig. 4.19).[20] The astigmatism of this objective was not noticeably different from that of previous designs, but the coma was actually worse, consisting of two orders of coma opposing each other, resulting in the presence of negative coma for rays near the middle of the aperture and positive coma for rays near the edge of the lens.

---

[19] Ger. Pat. 6,189 (1879). See also von Rohr, ref. 6, p. 304.
[20] U.S. Pats. 241,437–8 (1881). See also von Rohr, ref. 6, page 305.

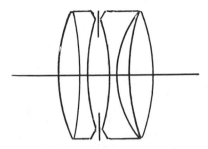

Figure 4.20. The Steinheil Rapid Antiplanet.

In his Antiplanet patent of 1881 Steinheil also disclosed his Portrait Antiplanet described in Chapter 3, Section IIIC. In 1893 Steinheil's son Rudolph designed another similar lens called the Rapid Antiplanet, an odd name considering that its aperture was only $f/6.3$.[21] The front component was similar to his father's Group Antiplanet, but the rear component was a novel cemented triplet consisting of a negative and a positive element of moderately high-index glass separated by a glass of much lower index (Fig. 4.20). This general type of construction was used later by R. Steinheil in his Orthostigmat design of 1893 (see Chapter 6, Section VB). Calculation shows that the astigmatism of the Rapid Antiplanet was somewhat less than in previous lenses but the coma was worse, and it is doubtful if the user would be able to tell the difference.

---

[21] Ger. Pat. 76,662 (1893). See also von Rohr, ref. 6, page 306.

CHAPTER 5

Optical Glass

I. EARLY HISTORY OF OPTICAL GLASS

At this point it is of interest to outline the historical development of optical glass, as this is the basic ingredient of all photographic lenses.

The earliest lenses were made from the crowns that appeared in window glass as a result of the glass-blowing processes in use at the time. These crowns were irregular lumps of glass much thicker than the rest of the pane; they may still be seen in old cottage windows in England. The glass was of indefinite and unknown composition and no two pieces were alike, although the differences were generally small. The glass suffered from many bubbles and striae, but it was all that the lens maker could obtain.

In the early 1700s when C. M. Hall and later J. Dollond, wished to make achromatic lenses, they needed a glass having higher dispersion than ordinary window glass, and they chose the flint glass commonly used to make fine tableware.[1] This name was given because broken flints were used as a source of silica instead of common sand, with the addition of lead oxide to

---

[1] H. C. King, *The History of the Telescope*, page 144. Sky Publishing Corporation, Cambridge, Mass., 1955.

improve the viscosity of the glass at the low red heat used in glass blowing. The presence of lead oxide also had the effect of raising the refractive index and dispersion of the glass, thus increasing the sparkle of cut tableware. The increased dispersion was just what was needed to make an achromatic lens.

By the end of the eighteenth century, efforts were being made to improve the glass used in lenses. In 1774 Pierre-Louis Guinand (1748–1824), a Swiss clockmaker, discovered how to remove the striae in glass by stirring the melt. He kept this process a secret for many years, but it failed to provide him with a livelihood. Eventually, in 1805, he was invited to join the German optical firm of Reichenbach, Utzschneider and Liebherr, who set him up as a glassmaker in an old monastery at Benedictbeurn in Bavaria, where he was expected to make whatever glass was needed by the firm.[2] In 1807 Joseph Fraunhofer (1787–1826), a twenty-year-old mirror maker was hired as a lens polisher and within two years he became a director of the firm.[3] In 1811, feeling that Guinand was not progressing very well, Fraunhofer took over the glassmaking and Guinand returned to Switzerland where he established his own glassworks at Les Brenets. On Guinand's death in 1824 his younger son Aimé (1774–1847) took over his father's workshop but let it go to ruin. Guinand's widow, Rosalie, in association with Theodore Daguet (1795–1870), founded another glassworks in France, but this also failed in 1857. Finally, in 1827, a year after Fraunhofer's death, Guinand's older son Henri (1771–1851), then serving as a clock maker in Clermont, also set up a glassworks in collaboration with the telescope maker J. N. Lerebours (1761–1840) and the glassmaker George Bontemps, at Choisi-le-roi. In 1832 he moved the factory to Paris where he was joined by his Swiss son-in-law Jean-Jacques Feil [originally Pfeil]. This company eventually became Parra-Mantois, which is still in existence in France under the name Sovirel.

In 1824, Robert Lucas Chance and his brother William purchased the British Crown Glass Company's works at Smethwick near Birmingham and in 1848 engaged the services of George Bontemps to introduce the manufacture of optical glass into England. The company flourished and until World War I was the only manufacturer of optical glass in the British

---

[2] W. H. S. Chance, "The Optical Glassworks at Benediktbeuern." *Proc. Phys. Soc.* **49,** 433 (1936).

[3] M. von Rohr, *Joseph Fraunhofers Leben, Leistungen, und Werksamkeit,* page 102. Akademische Verlagsgesellschaft, Leipzig, 1929. A partial listing of Fraunhofer's glass types is given in von Rohr's *Theorie und Geschichte des photographischen Objektivs,* page 329. Springer, Berlin, 1899.

Empire. In 1932 Chance absorbed the Parsons Optical Glass Company, which had been established in 1917 as a branch of the Derby Crown Glass Company and was acquired by Charles Parsons in 1921. Chance was finally united with Pilkington to form the Chance-Pilkington Optical Glass Works at St. Asaph, where it is now a very active establishment.

In the United States, optical glass was first made on an experimental scale by the Bausch & Lomb Optical Company in Rochester in 1912. This proved to be an extremely fortunate circumstance when all supplies of European glass were abruptly cut off by the War of 1914 and the United States had to depend entirely on domestic products. A glass factory was immediately established by Bausch & Lomb and by 1917 they were making large quantities of barium crowns suitable for the manufacture of photographic objectives.[4] Their first optical glass catalog was issued in 1919 containing 25 different types of glass. Several other optical glassworks were established during the first World War, including one at Hamburg, New York, and another at the National Bureau of Standards in Washington, D.C. During the second World War enormous quantities of optical glass were made by Corning and the Pittsburgh Plate Glass Company.

II. OPTICAL PROPERTIES OF GLASS

This might be a good place to explain the optical properties of glass that serve to distinguish optical glass from the more common types such as those used to make windows and bottles.

The refractive index of all glasses increases toward the blue end of the spectrum and when Fraunhofer made the first systematic attempts to classify glasses he used the dark lines in the solar spectrum as fiducial marks to indicate wavelengths. In this way, he used the yellow sodium D line to express the refractive index of the glass and he used the red C line and the blue F line of hydrogen to define the dispersion. Thus

$$\text{mean dispersion} = (n_F - n_C),$$

$$\text{dispersive power} = \frac{n_F - n_C}{n_D - 1}.$$

[4] F. E. Wright, *The Manufacture of Optical Glass and of Optical Systems,* U.S. Ordnance Department Document 2037. Government Printing Office, Washington, D.C., 1921.

Abbe disliked the smallness of the numerical value of the dispersive power of glass, so he used the reciprocal of the dispersive power, a number ranging from about 25 to 70, instead. This value is now called the V-value, or Abbe Number, of the glass. Unfortunately, a high dispersive power is represented by a low V-value and vice versa, but opticians soon become familiar with this anomaly. It can readily be shown that the chromatic aberration of a simple lens is equal to the focal length divided by the V-number of the glass and that an achromatic lens can be made by combining a strong element of crown glass with a weaker element of flint glass, the division between crowns and flints being arbitrarily set at a V-number of 50.

III. THE SCHOTT GLASS WORKS

It was explained in Chapter 1, Section IB that in order to make a lens having a flat field free from astigmatism it is necessary to reduce the Petzval sum, and that one way to do this is to use a pair of glasses in which the crown has a high refractive index and the flint a low index. This relation is contrary to the old arrangement in which flint glasses have a higher refractive index than crown glasses.

The first successful high-index crown glasses were manufactured by Abbe and Schott at the Jena Glassworks in Germany.[5] Ernst Abbe (1840–1905), a young 26-year-old physics professor at the University of Jena, was hired in 1866 by Carl Zeiss (1816–1888) to put his instrument workshop on a proper scientific basis. Abbe worked first on the design of the microscope and by 1880 he realized that he needed some radically new types of glass to remove the secondary spectrum of a microscope objective. So he persuaded the 29-year-old Otto Schott (1851–1935), a glassmaker from Witten, to join him in establishing a glass factory at Jena.

In the incredibly short time of six years they were able to issue a catalog containing 44 types of glass, many of which were entirely novel. In Figure 5.1 is shown a plot of the refractive indices of their glasses against the mean dispersion ($n_F - n_C$), and in Figure 5.2 there is a similar plot of refractive index against the V-number of the glass. At the low-dispersion end (high V-number) were four phosphate crowns, which were interesting but unfortunately so unstable chemically that they had to be withdrawn. Next came three barium crowns, useful in reducing the Petzval sum, followed by a series of borate flints, which were the glasses Abbe was seeking to reduce the secondary spectrum of a microscope objective.

---

[5] H. Hovestadt, *Jena Glass*, trans. J. D. Everett. Macmillan, New York, 1902.

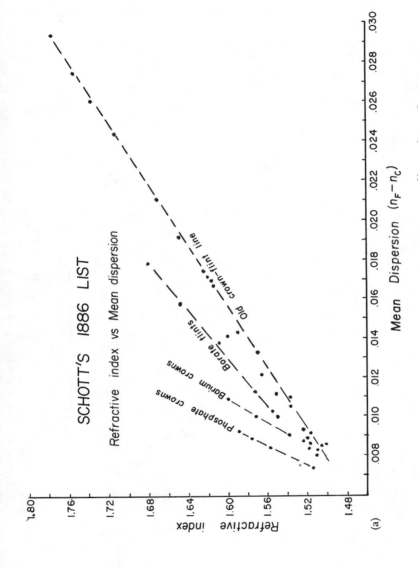

Figure 5.1. (a) Plot of rafractive index vs. mean dispersion of original Schott glasses.

73

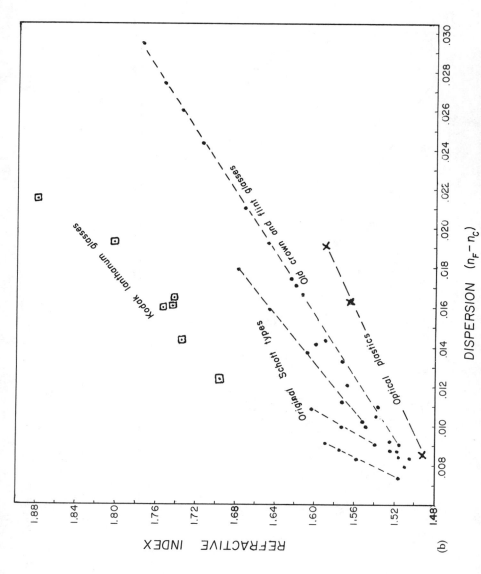

Figure 5.1. (b) The same as Figure 5.1(a) with the addition of early lanthanum glasses and plastic materials.

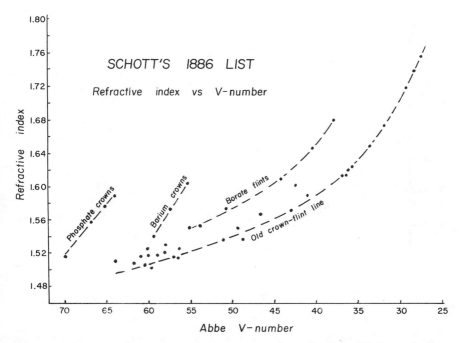

Figure 5.2. Plot of refractive index vs. V-number of original Schott glasses.

Secondary spectrum is caused by a difference between the partial dispersion ratio of the crown and flint glasses used in an achromat. By partial dispersion is meant the value of $n_{G'} - n_F$ at the blue end of the spectrum, usually expressed as a fraction of the mean dispersion; thus,

$$P_{G'F} = (n_{G'} - n_F)/(n_F - n_C).$$

In Figure 5.3 is shown a plot of the partial dispersion ratio $P_{G'F}$ against the V-number of the glasses in the original Schott list, and it is clear that the borate flints are significantly different from the old crown-flint glasses in this regard. These glasses were useful to Abbe, but they too were chemically unstable and were eventually replaced by more stable types.

The major success of the new glassworks was the development of barium crowns. An achromat composed of a high-index crown element combined with a low-index flint is known as a new achromat.[6] Its Petzval sum is small,

---

[6] O. Lummer, *Contributions to Photographic Optics*, trans. S. P. Thompson, page 47. Macmillan, London, 1900.

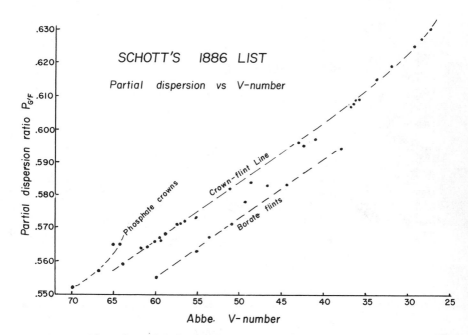

Figure 5.3. Plot of partial dispersion ratio vs. V-number of original Schott glasses.

which was the reason for its use, but the cemented interface is collective instead of dispersive, so the spherical aberration of the doublet is greater than in an old achromat, and the field tends to be inward-curving, just the opposite of the Chevalier landscape lens. During the following years the number of barium crown glasses in the Schott catalog was steadily increased, and these glasses were immediately adopted by lens designers in an effort to reduce the astigmatism that had plagued all previous photographic lenses.

After World War II, Jena came into the Russian zone of Germany, and the staffs of the Zeiss Company and the Schott Glassworks were split into two. One part of each decided to remain at Jena in the Russian zone, while the rest of Zeiss was established in the little town of Oberkochen in the American zone, and the rest of Schott was moved into Mainz in 1952. In 1967 the demand for optical glass from America was so great that Schott decided to establish a branch factory at Duryea in Pennsylvania. The current Schott catalog is a veritable textbook on the properties of optical glasses. It contains data on hundreds of types, from the lightest crowns to the densest lanthanum glasses and the highly dispersive titanium flints, all of which have their uses in optical technology.

IV. THE LANTHANUM CROWNS

During the 1930's, G. W. Morey of the Geophysical Laboratory in Washington, D.C., undertook to develop new and unusual optical glasses, mainly aimed at increasing still further the high refractive index of the barium crowns. He discovered in 1934 that a borate glass containing the rare earth element lanthanum was very favorable, and he also included elements such a thorium and tantalum in his experimental glasses. In 1937 the Eastman Kodak Company undertook to develop some of Morey's novel glasses, which had to be melted in a platinum crucible, and in a short time they had worked out seven lanthanum crowns, indicated on the chart in Figure 5.1(b).[7] Today, every optical glass manufacturer is making lanthanum crowns and flints, and the availability of these extreme glasses has greatly helped to improve photographic objectives. Current lens patents indicate that lanthanum glasses are used in practically every modern photographic objective.

V. OPTICAL PLASTICS

Plastic materials such as methyl methacrylate (Lucite and Plexiglas in the United States; Perspex in England) were developed in the 1920s, and it was early proposed that they could be used to make unbreakable eyeglass lenses.[8] This is now a thriving business. However, it was then not considered possible to make plastic lenses that would be good enough for photographic objectives. But progress was made and in 1936 a company called Combined Optical Industries was established in England to make precision lenses. In 1952 the Eastman Kodak Company began making plastic viewfinder lenses, and these were so satisfactory that starting in 1957 they began to make the taking lenses for low-cost cameras from plastic mold-

[7] R. Kingslake and P. F. DePaolis, "New Optical Glasses." *Nature* **163**, 412 (1949); also *Scientific Monthly* **68**, 420 (1949).

[8] H. C. Raine, "Plastic Glasses," *in* Optical Instruments: Proceedings of the London Conference 1950 (W. D. Wright, ed.), page 243. Chapman & Hall, London, 1951.

ings. Well over 100 million plastic camera lenses have been made since that time.

The virtues of plastic lenses are many. They can be made at low cost by injection molding of the melted material into a polished mold and, of course, it is just as easy to make aspheric lenses as spheres once the aspheric mold has been fabricated. Furthermore, it is possible to form a mount around the lens, either as a mechanical protection against abrasion or so that two or more lenses can be assembled together without the need for any metal rings or clamps. For example, in Figure 5.4 is shown a plastic triplet lens made by Kodak; the front and rear elements are equipped with flat flanges for mounting into a metal plate, while the inner negative element, formed from a different material, is cemented into a recess behind the front element. Molded plastic lenses show a greater uniformity in thickness than the corresponding glass elements. Plastic materials are light in weight and some very large lenses are made of plastic for this reason.

The disadvantages of plastic materials are also numerous, but the advantages predominate. There are only three or four plastic materials that are suitable for lens manufacture; they are indicated at the bottom of the chart in Figure 5.1(b). There are no plastics corresponding to the barium crown glasses, so the designer must make do with only low-index materials.

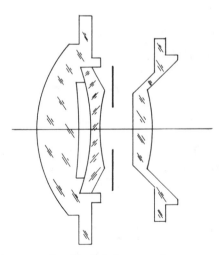

Figure 5.4. The lens on the Kodak Starmatic camera (molded plastic).

Plastics have a high temperature coefficient of refractive index, some 120 times that of glass, so special precautions must be taken to prevent the camera from going out of focus at high or low ambient temperatures. Another manufacturing problem is that plastic lenses tend to acquire a static charge that attracts dust, which cannot be removed without risk of scratching the surface. However, these difficulties are being overcome, and the prospects for plastic lenses are very good indeed. At present it is common to find that plastic lens elements are used to correct aberrations, with most or all of the lens power being in a glass element that has a negligible temperature coefficient of refractive index.

VI. OTHER MATERIALS

Besides optical glass and plastics, a few other materials have been used to make lenses, principally crystalline and fused quartz, crystals such a calcium and lithium fluoride for the ultraviolet, and a number of special materials such as silicon and germanium for the infrared. Many of the latter materials are opaque in the visible but quite transparent at wavelengths beyond one or two microns.

Sometimes these infrared materials have been ground into a fine powder and then hot-pressed into a mold approximating the size and shape of the finished lens. This procedure cannot be used in the visible because the grain boundaries scatter blue light, but this is no problem in the longer wavelengths.

Some lenses have been manufactured recently in which crystalline fluorite has been used for the crown components and a barium crown glass for the flint components. The reason for this is that the partial dispersion ratio of fluorite is similar to that of barium crown glass but the refractive index and V-number are different. Such a combination exhibits no secondary spectrum and all colors come to a common focus. The same arrangement has been used for a long time in apochromatic microscope objectives. The secondary spectrum of short-focus camera lenses is generally insignificant, but it can be quite serious in a long-focus telephoto.

Occasionally it has been proposed to make a lens from a pair of suitably shaped plastic shells, which can be of any desired spherical or aspherical shape, cemented together and filled with water or oil. The risk of leakage and the inevitable presence of thermal gradients in the liquid make such lenses suitable only for condensers or field lenses. Some very large liquid lenses have been made where a solid material would be impractical.

REFERENCES

G. W. Morey, "The Properties of Glass." Reinhold, New York, 1938.

H. W. Lee, "The New Optical Glasses." *Sci. Prog.* **34,** 533 (1946).

I. C. Gardner, "New Types of Optical Glass Available in the United States." See Footnote 8, page 241.

D. F. Horne, "Optical Production Technology," page 443. Crane Russak, New York, 1972.

A. J. Marker, "Optical Glass Technology." *Proc. SPIE* **531,** Paper 01 (1985).

S. Musikant, "Optical Materials." Marcel Decker, New York, 1985.

CHAPTER 6

The First Anastigmats

I. NEW-ACHROMAT DOUBLETS

It was explained on page 75 that a new-achromat cemented doublet made with a high-index crown glass and a low-index flint has a low Petzval sum, and thus offers the possibility of making a flat-field lens free from astigmatism.

As soon as barium crown glasses became available that met this condition, H. L. H. Schroeder, then working at Ross in England, developed a symmetrical doublet consisting of two identical new-achromat menisci arranged about a central stop (Fig. 6.1).[1] Because the spherical aberration was uncorrected, the aperture of this lens was limited to about $f/20$, but the image over the entire field of $\pm 30°$ was indistinguishable from the axial image, the lens showing no signs of either field curvature or astigmatism. Such perfection has never been attained before. This lens, known as the Concentric, was marketed by Ross for many years.

---

[1] Brit. Pat. 5,194/88; U.S. Pat. 404,506.

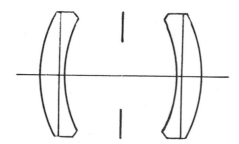

Figure 6.1. The Ross Concentric lens.

II. THE ZEISS ANASTIGMATS

The low aperture of the Concentric was a severe limitation, and in 1890 Paul Rudolph (1858–1935) of Zeiss succeeded in correcting the spherical aberration of a new-achromat landscape lens by the addition of a low-power aberration corrector in front of the stop (Fig. 6.2). Rudolph regarded the result as a combination of an old-achromat front component with a new-achromat rear component.[2] He called this lens the Anastigmat, and from 1890 to 1893 various modifications appeared on the market:

1890	Series III	Aperture $f/7.2$	Cemented doublet in the rear.
	Series IV	Aperture $f/12.5$	Cemented doublet in the rear.
	Series V	Aperture $f/18$	Cemented doublet in the rear.
1891	Series I	Aperture $f/4.5$	Cemented triplet in the rear.
	Series II	Aperture $f/6.3$	Cemented triplet in the rear.
	Series IIIa	Aperture $f/9$	Cemented doublet in the rear.
1893	Series IIa	Aperture $f/8$	Cemented triplet in the rear.

The Series I lens with an aperture of $f/4.5$ was intended for portraiture, and the Series V $f/18$ lens covered an exceptionally wide angle.

The word Anastigmat is actually a double negative. The Greek word stigma means a point and, of course, the image of a point should be a point. The word A-stigmatism means a no-point, or imperfect definition, and An-a-stigmat means a no-no-point, and hence good definition. An image

---

[2] Ger. Pat. 56,109; U.S. Pat. 444,714; Brit. Pat. 6,028/90.

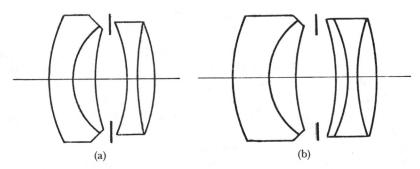

Figure 6.2. Zeiss Anastigmats with (a) two elements and (b) three elements in the rear component.

free from astigmatism is said to be Stigmatic, and only if this image falls on a flat plane is it called Anastigmatic.

Zeiss licensed several foreign companies to manufacture their Anastigmat lenses, including:

Bausch and Lomb, Rochester, New York
Krauss, Paris, France
Ross, London, England
Fritsch (formerly Prokesch), Vienna, Austria
Koristka, Milan, Italy
Suter, Basle, Switzerland.

By 1900 Zeiss claimed that they and their licensees had sold over 100,000 Anastigmats.

In spite of several redesigns, the Anastigmats did not prove to be as good as was hoped. This was partly due to the small number of glass types available at that time, but it was more likely due to the higher-order aberrations introduced by the strong cemented interface in the front component. Consequently, when the much better Planar, Unar, and Tessar lenses appeared on the market, most of the Anastigmats were withdrawn. In their 1901 catalog Zeiss listed only the Series IIa, IIIa, and V Anastigmats, and by 1910 only the Series IIIa and V remained. Zeiss also lost their copyright on the name Anastigmat when other companies began calling their lenses anastigmats, so in 1900 Zeiss changed the name of their lenses to Protar. The $f/18$ Series V Protar survived as the standard wide-angle lens for commercial photography up to the 1930s.

III. THE ALDIS STIGMATIC LENSES

In 1895 Hugh L. Aldis, secretary to the firm of Dallmeyer Ltd., revived
the plan of the Zeiss Anastigmats in a series of novel lenses called the
Stigmatics. It will be recalled that in the Zeiss Anastigmats the front component was an old achromat of almost zero power with a strong dispersive
interface for the correction of spherical aberration, while the rear component was a new achromat carrying almost all the power of the system. In
the Stigmatic lenses, Aldis replaced the strong dispersive interface in the
front component by a narrow airspace having the shape of a positive lens.
This airspace has the same effect as the dispersive internal surface so far as
spherical aberration is concerned, but with better zonal correction. The
Zeiss designer Paul Rudolph used the same device in his later Unar and
Tessar designs.

The Stigmatic lenses (Fig. 6.3) were disclosed in a British patent.[3] A
positive component carrying most of the power was placed on one side of

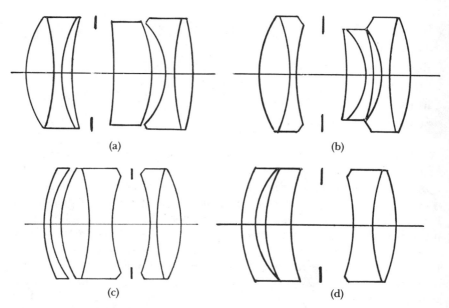

Figure 6.3. The Aldis Stigmatic lenses: (a) original $f/3.5$; (b) Series I, $f/4$;
(c) Series II, $f/6$; and (d) Series III, $f/7.5$.

---

[3] Brit. Pat. 16,640/95; U.S. Pat. 560,460; Ger. Pat. 92,582.

the stop, with a pair of closely-spaced positive and negative menisci on the other side, the space between them having the shape of a positive meniscus lens. The inventor noted that this arrangement could be used either way round.

In the original form, announced as an $f/3.5$ portrait lens (Fig. 6.3(a)), the front component was a cemented triplet, one cemented interface being used for spherical correction and the other for chromatic correction. The two components in the rear were a single positive meniscus and a negative meniscus achromat. In the Series I Stigmatic the spherical-correcting interface was moved from the front to the middle lens, and the aperture was reduced to $f/4$ (Fig. 6.3(b)). All the positive elements in this system were made of the same medium barium crown glass, the Petzval sum was about 0.03 on a focal length of 10. The Series I design was offered for sale by Dallmeyer for over ten years.

In the Series II $f/6$ Stigmatic, (Fig. 6.3(c)), which covered a field of about $\pm 30°$, it appears that Aldis began to realize that it was not necessary to achromatize each of the three components separately, and he therefore turned the whole system around end-for-end and replaced the negative doublet by a simple negative meniscus element. The middle and rear components were almost identical new-achromat doublets. It was claimed that the system was convertible so that each half could be used alone. This lens was offered for sale until at least 1911.

In the Series III $f/7.5$ Stigmatic (Fig. 6.3(d)) the designer went further and made each of the front components into a single element, negative in front and positive behind, in edge contact with the usual positive airspace between them. This arrangement was not separable.

In 1910, this lens was redesigned to make the low-cost $f/6.3$ Series IV Stigmatic lens, also called the Carfac. It was offered at half the price of the Series II lens and it was claimed to be convertible, the front component

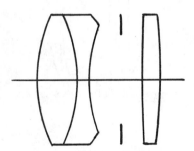

Figure 6.4. The Aldis lens.

having 3 times and the rear component 1½ times the focal length of the combined system.

In 1901, when H. L. Aldis had left Dallmeyer and established his own company, he simplified the lens still further by the use of a thick cemented doublet of low power in front and a biconvex positive element behind (Fig. 6.4).[4] This lens might be regarded as a Triplet in which the two front elements had been thickened and cemented together. These Aldis lenses were sold in various forms and apertures until about 1920.

IV. THE UNAR AND TESSAR

It was mentioned in the last section that H. L. Aldis, in 1895, employed a narrow airgap in the shape of a positive lens to correct the spherical aberration in his Stigmatic objective. This device is much better than using a strong dispersive cemented interface for this purpose, as the resulting zonal spherical aberration is much smaller. A further incidental advantage of the use of an airspace is that designers are then much freer in their choice of glasses, because they may use glasses having the same refractive index on both sides of the airspace, a procedure that would be useless in a cemented lens.

In 1899, Paul Rudolph realized this, and he proceeded to replace both of the cemented interfaces in his Anastigmat design by narrow airspaces and thus produced his Unar lens, shown in Figure 6.5.[5]

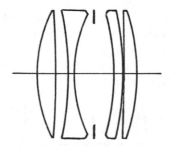

Figure 6.5. The Zeiss Unar.

---

[4] Brit. Pat. 5,170/01; U.S. Pat. 682,017.
[5] Ger. Pat. 134,408; U.S. Pat. 660,202; Brit. Pat. 24,089/99.

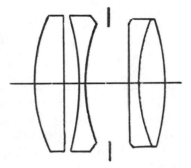

Figure 6.6. The Zeiss Tessar.

Although the Unar was a good design, Rudolph found that the strong collective interface in the rear component of the Anastigmat had many virtues, so he restored the rear cemented doublet of the Anastigmat while retaining the airspaced front component of the Unar and thus produced his wonderfully successful Tessar design of 1902 (Fig. 6.6). The front component of the Tessar, like that of the Anastigmat, had very little power, its sole function being to correct the remaining aberrations of the strong new-achromat rear component. The collective cemented interface in the rear component serves three functions: 1) it reduces the zonal spherical aberration; 2) it reduces the overcorrected oblique spherical aberration, and 3) it reduces the gap between the astigmatic foci at intermediate field angles.

The Tessar first appeared at an aperture of $f/6.3$,[6] but by 1917 the aperture had been raised to $f/4.5$, and finally to $f/2.8$ by W. Merté and E. Wandersleb in 1930.[7] Lenses of the Tessar type have been made by the millions, by Zeiss and every other manufacturer, and they are still being produced as an excellent lens of intermediate aperture. The famous 50 mm $f/3.5$ Elmar lens fitted to the early Leica cameras was of this type, designed by Max Berek (1886–1949) in 1920.[8]

Actually, Zeiss enjoyed a long monopoly on this type of construction, as Rudolph's patent was extremely general. His single claim reads:

"A spherically, chromatically, and astigmatically corrected objective consisting of four lenses separated by the diaphragm into two groups

---

[6] Ger. Pat. 142,294; U.S. Pat. 721,240; Brit. Pat. 13,061/02.
[7] Ger. Pat. 603,325; U.S. Pat. 1,849,681; Brit. Pat. 369,833 (1930).
[8] Ger. Pat. 343,086 (1920).

each of two lenses, of which groups one includes a pair of facing surfaces and the other a cemented surface, the power of the pair of facing surfaces being negative and that of the cemented surface positive."

Soon after this patent issued, it was pointed out to Rudolph that the Series III Stigmatic of H. L. Aldis came under this claim, although it had never been patented. Consequently Zeiss issued a codicil stating that the positive element of the separated lenses must be outside and the negative element inside toward the stop. This was, of course, the opposite of the arrangement used by Aldis in the Stigmatic objective.

The American Tessar patent appeared in February 1903, thus giving Zeiss a monopoly on this design until 1920, at the end of World War I. Some of the more notable lenses of the Tessar type appeared under the following trade names, although not every lens with one of these names is necessarily of this construction:

Agfa: Solinar
Berthiot: Flor, Olor
Boyer: Saphir
Busch: Glyptar
Dallmeyer: Dalmac, Perfac, Serrac
Ernemann: Ernon
Hermagis: Hellor, Lynx
Ilex: Paragon
Kodak: Ektar
Laack: Dialytar
Leitz: Elmar, Varob
Meyer: Primotar
Plaubel: Anticomar
Rodenstock: Ysar
Ross: Xtralux
Roussel: Stylor
Schneider: Comparon, Xenar
Taylor-Hobson: Apotal, Ental
Voigtländer: Heliostigmat, Skopar
Wollensak: Raptar
Wray: Lustrar

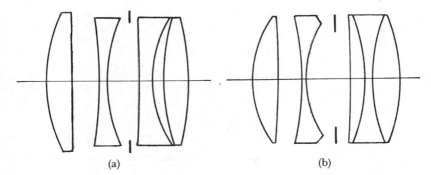

Figure 6.7. Variations on the Tessar: (a) the Ross Xpres and (b) the Berthiot Olor.

A few designers have turned the lens around so that the cemented doublet comes in front and the separated elements behind. Among these designs were:

Agfa: Solinar
Isco: Westanar
Meyer: Primotar
Steinheil: Culminar, Neodar, Triplar
Voigtländer: Avus, Heliostigmat.

A. Modifications of the Tessar

The Tessar was such an excellent design that other workers would have liked to copy it but were prevented from doing so by patent limitations. The simplest way out was to use a cemented triplet in the rear instead of a doublet. Several designs of this type appeared in 1913, including the Ross Xpres by J. Stuart and J. W. Hasselkus,[9] the Gundlach Radar lens, and the Berthiot Olor and related lenses by Florian.[10] These designs were similar (Fig. 6.7). In 1925 E. Wandersleb and W. Merté of Zeiss designed the $f/2.8$ Biotessar, consisting of a cemented doublet in front, a single negative element, and a cemented triplet behind (Fig. 6.8).[11]

It is possible to regard the Voigtländer Heliar and its variations as being modifications of the Tessar, but the Heliar appeared before the announce-

[9] Brit. Pat. 29,637/13.
[10] U.S. Pat. 1,122,895 (1913).
[11] Ger. Pat. 451,194; U.S. Pat. 1,697,670; Brit. Pat. 256,586 (1925).

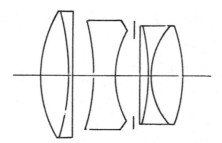

Figure 6.8. The Zeiss Biotessar.

ment of the Tessar. Actually, it was quite clearly a modification of the Triplet (see Chapter 7, II). It is certain that the Tessar was *not* a modified Triplet, as the series of steps followed by Rudolph in going from the Anastigmat to the Tessar are well established, but for some of the later designs it is not always clear whether they should be regarded as modified Tessars or modified Triplets.

V. SYMMETRICAL ANASTIGMATS
A. The Dagor Lens

In 1892, two years after the announcement of the Zeiss Anastigmat, Emil von Höegh (1865–1915) a 27-year-old mathematician, designed privately a cemented triple lens that was corrected for spherical and chromatic aberration and had a flat field free from astigmatism. Because it was not corrected for coma, this lens had to be assembled in pairs about a central stop so that the symmetry would remove the coma automatically. The story is told that von Höegh offered the design to Zeiss, but they were not interested, so he took it to Goerz in Berlin.[12] At that time the Goerz Company was only four years old, making a line of rectilinear lenses known as the Lynkeioskop, designed by Carl Moser (1858–1892). Goerz fabricated a sample of von Höegh's lens, which proved to be excellent, so they offered him a position as their principal lens designer in succession to Moser, who had recently died. The new lens was offered for sale as the Double Anastigmat Goerz, a name that was reduced to the acronym Dagor in 1904. It is still made by Goerz in the United States, and the same type has been manufactured by practically every lens company.

---

[12] R. Schwalberg, "Dagor: The Lens That's 78 Years Young." *Pop. Phot.* **70,** 56 (January 1972).

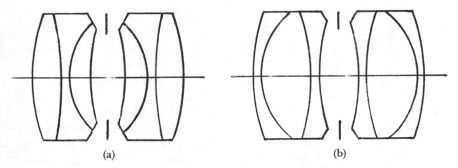

(a) (b)

Figure 6.9. (a) The original Dagor and (b) the reversed Dagor.

The Goerz patent was remarkably simple, as it claimed merely a cemented triple meniscus system in which the refractive index of the three elements increased steadily from one end to the other, the middle element being either positive or negative (Fig. 6.9).[13] The design of the original Dagor is shown in Figure 6.9(a). Within two months of Goerz's patent application, Rudolph of Zeiss applied for a practically identical patent in England, but it was never issued, probably because von Höegh's German patent had been published by then. Nevertheless, Zeiss did make a number of lenses named the Satz Anastigmat Series VI that were almost identical with the Goerz Dagor, but this lens was soon replaced by the quadruple Anastigmat Series VII.

The design of a Dagor lens is not difficult provided that there is a sufficiently wide range of glass types available. The designer selects a series of three likely refractive indices for which a broad range of dispersive powers exists and solves for the outside radii of curvature to give the desired focal length and Petzval sum. Then one of the internal cemented interfaces is used to correct the spherical aberration, while the other is chosen to flatten the field, the actual glass choice being left to the end for chromatic correction. The three transverse aberrations, coma, distortion, and lateral color, are automatically eliminated by the use of a symmetrical combination of two identical lenses about a central stop. Von Höegh must have had some trouble with achromatization as there were very few barium crown glasses available at that time.

The new Goerz lens was an immediate success, and we are told that by 1895 some 30,000 had been sold. The lens covered a fairly wide field ($\pm 30°$ or more) at a moderate aperture ($f/6.8$). However, it did suffer

---

[13] Ger. Pat. 74,437; U.S. Pat. 528,155; Brit. Pat. 23,378/92.

from a serious amount of zonal spherical aberration, which arose at the strong dispersive interface, with the result that the position of sharpest image tended to move along the axis when the lens was stopped down, a phenomenon known as focus shift. Nevertheless, the lens has always been popular with photographers, both professional and amateur.

Some designers have preferred the reversed Dagor type of construction shown in Figure 6.9(b). The two designs shown here to scale are comparable in every way. The focal length, F-number, Petzval sum, astigmatism, and spherical aberration are the same in both, and the radii of curvature, zonal aberration, and field curves are very similar. The only significant difference between the two designs is the refractive index of the internal element, which must be low if it is negative and high if it is positive. In these two examples the glasses were:

Elements	Lens A (Dagor)		Lens B (reversed Dagor)	
1 and 6	$n = 1.617$	$V = 56$	$n = 1.617$	$V = 56$
2 and 5	$n = 1.547$	$V = 51$	$n = 1.590$	$V = 58$
3 and 4	$n = 1.517$	$V = 54$	$n = 1.517$	$V = 52$

although a broad range of glass selection is possible in both designs. In the original Dagor the two outer elements are both positive, while in the reversed Dagor they are both negative. The reversed type is somewhat thicker than the normal Dagor, leading to the risk of excessive vignetting. This arrangement was used by Zeiss in their Ortho-Protar and also by A. E. Conrady in the Watson Holostigmat lens of 1905. The vignetting problem was avoided in the wide-angle Schneider Angulon lens of 1930 by making the outer surfaces decidedly larger than the diameter of the axial beam (Fig. 6.10).[14]

B. The Orthostigmat and Collinear

Simultaneous inventions often occur in the field of lens design. We have already referred to the simultaneous invention of the Rapid Rectilinear and the Aplanat of 1866 and also to the simultaneous invention of the triple cemented anastigmats of von Höegh and Rudolph in 1892. In November of 1893, R. Steinheil in Munich patented a lens called the Ortho-

---

[14] Ger. Pat. 579,788; U.S. Pat. 1,882,530 (1930).

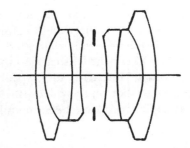

Figure 6.10. The Schneider Angulon.

stigmat,[15] which was similar to the Goerz Double Anastigmat except that the elements were in a different order (Fig. 6.11). As in the Goerz design, each half of the Orthostigmat contained a strong dispersive surface to correct the spherical aberration and a weaker collective surface to flatten the field, but in the Orthostigmat both surfaces were convex towards the stop. In his patent Steinheil claimed a positive element cemented between a biconvex element and a biconcave element, both outer elements having a higher refractive index than the enclosed positive element.

Two years later, in July 1895, F. R. von Voigtländer patented a virtually identical lens called the Collinear in which he claimed a low-index element cemented between two other elements, one of the outer elements being of intermediate refractive index and the other of high refractive index.[16] It is hard to understand how Voigtländer could obtain a patent on a lens so

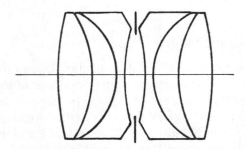

Figure 6.11. The Voigtländer Collinear and the Steinheil Orthostigmat.

---

[15] Ger. Pat. 88,505 (1893); Brit. Pat. 12,949/95.
[16] U.S. Pat. 567,326 (1895).

close to Steinheil's, but it is possible that his designers H. Scheffler and D. Kaempfer had been working on this design before Steinheil's application date. Nevertheless, both lenses were marketed and both were excellent. In some instances the Collinear lens carried the number of the Orthostigmat patent engraved on the barrel, so it appears that the two companies worked very closely together on these lenses. Many objectives of the symmetrical cemented triplet type have been manufactured, including:

Boyer: Beryl
Busch: Leukar
Ernemann: Ernon
Goerz: Aerotar, Geodar
Hermagis: Aplanastigmat
Laack: Polyxentar
Meyer: Silesar
Plaubel: Makinar, Triple Orthar
Reichert: Neucombinar
Rodenstock: Eikonar, Teragonal
Roussel: Antispectroscopique
Schneider: Angulon
Simon: Hexanar
SOM: Eurygraphe, Perigraphe
Staeble: Neoplast, Protoplast
Watson: Holostigmat
Zeiss: Amatar

C. The Airspaced Dagor Type

The large zonal spherical aberration of the Dagor can be drastically reduced by replacing the strong cemented interface of the Dagor by an airspace having the shape of a convex lens. (The same device led Rudolph to develop the Unar and Tessar from his earlier Anastigmat.) In 1903 E. Arbeit of the firm of Schultz and Billerbeck of Potsdam patented the Euryplan lens of the airspaced Dagor type (Fig. 6.12), in which the inner positive elements were separated from the remaining cemented doublets.[17] This gave designers several additional degrees of freedom in which

---

[17] Ger. Pat. 135,742; Brit. Pat. 2,305/03.

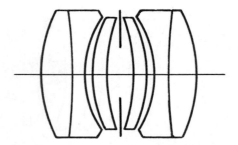

Figure 6.12. The separated Dagor type, the Plasmat.

to work, permitting them to raise the aperture to $f/4.5$ and increase the field at the same time. The presence of eight glass-air surfaces caused some problems because of interreflections between the surfaces, as was mentioned earlier in Chapter 1, Section VB. Nevertheless, this type has proved to be quite satisfactory in practice and it has been used extensively by several manufacturers.

Prior to World War I it was the custom in some companies to pay designers a royalty on every one of their lenses that was sold. By 1911 Paul Rudolph was able to retire on his earnings and live the life of a country gentleman. Unfortunately, he suffered such a financial loss in the inflation that followed the war that he had to go back to work at the age of 61. He returned briefly to Zeiss, but realized that he would much rather work in a smaller company, so he accepted employment with the firm of Hugo Meyer, which had been founded in Görlitz in 1895. There he designed a series of lenses known generally as Plasmats. His first design was the $f/4.5$ Double or Satz Plasmat of the airspaced Dagor type shown in Fig. 6.12.[18] Meyer also made a similar lens called the Euryplan, so they must have acquired rights to use this name from Schultz and Billerbeck. It is said that the name Plasmat was given to these lenses because they exhibited more depth of field than usual; however, this seems unlikely, as an increase in depth can be obtained only by the deliberate introduction of excessive spherical or chromatic aberration (see Chapter 13, IF).

The same type of construction was used by Zeiss in their unsymmetrical Orthometar lens designed by W. Merté in 1926.[19] This and the similar Ross Wide-angle Xpres[20] were regarded as the standard lenses for aerial

[18] Ger. Pat. 310,615; Brit. Pat. 135,853 (1918).
[19] Ger. Pat. 649,112; U.S. Pat. 1,792,917; Brit. Pat. 278,338 (1926).
[20] Brit. Pat. 295,519: U.S. Pat. 1,777,262 (1927).

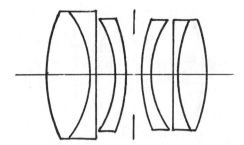

Figure 6.13. The Meyer Kino Plasmat.

photography for many years, as they covered a field of ± 35° with negligible distortion; they were finally replaced by the Topogon and Metrogon about 1932. Symmetrical lenses of the Plasmat type have been used extensively in document copying machines. They have occasionally been used as ordinary camera lenses, including:

Bausch & Lomb: Animar
Boyer: Saphir B
Reichert: Polar
Schneider: Symmar
SOM: Orthar

About 1922 Rudolph designed an $f/2$ and $f/1.5$ lens called the Kino Plasmat (Fig. 6.13). This lens was almost symmetrical, the inner negative elements being convex toward the stop.[21] This orientation of the negative elements is highly effective in correcting the spherical aberration, but it leads to a limited angular field. However, this was no problem in lenses intended for use on a motion-picture camera where the field must be narrow for reasons of perspective.

D. Quadruple Cemented Lenses

The first cemented quadruplet lens was the Double (or Quadruple) Anastigmat Series VII, designed by Paul Rudolph in 1894[22] and manufactured extensively by Zeiss. It was intended to be convertible (i.e., the rear component could be used alone or in combination with a similar front

---

[21] Ger. Pat. 401,630; U.S. Pat. 1,565,205 (1922).
[22] U.S. Pat. 532,398; Brit. Pat. 19,509/94.

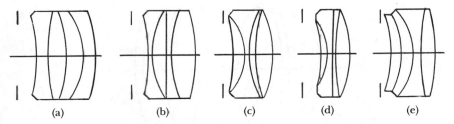

Figure 6.14. Five arrangements of a quadruple cemented objective.

component of the same or of a different focal length). For example, in Zeiss's Set No. C, the three components had focal lengths of 22, 28, and 35 cm, and with them one could construct double objectives having focal lengths of 14, 15½, and 18 cm, thus providing the user with six focal lengths ranging from 14 to 35 cm, obviously a great convenience for the practical photographer. The single rear components worked at $f/12.5$ and the double combinations at about $f/6.3$. Each half of this lens could be regarded as a new achromat located close to the stop, cemented to an old achromat on the outside, thus resembling the two components of the original Anastigmat arrangement. The use of four elements permitted better correction of coma than would be possible in a triplet, and hence the rear of a quadruplet could be used alone as an anastigmatic landscape lens, which was not possible with the Dagor.

Several different arrangements of four elements cemented together have been used, as indicated in Figure 6.14. Arrangement (a), containing elements in order outward from the stop of $(-++-)$, was the one used by Zeiss in 1894. It has also been used by other manufacturers in such lenses as the Beck Bistigmat, the Ross Combinable,[23] and the Wollensak Raptar Convertible. Arrangement (b), in order $(-+-+)$, was used in the Rietzschel Linear of 1898.[24] This system can be regarded as the Steinheil Orthostigmat with a plane interface inserted into the middle element.

Combination (c), in order $(+-++)$, was used in many quadruplet lenses including the Bausch & Lomb Plastigmat designed by Edward Bausch in 1900,[25] the Reichert Combinar designed by Heimstädt in 1902,[26] the Simon Octanar of 1902,[27] the Goerz Pantoplan, and the Gundlach Verastigmat of 1907.

[23] Brit. Pat. 29,636/13.
[24] Ger. Pat. 118,466; Brit. Pat. 6,080/01.
[25] U.S. Pat. 660,747 (1900).
[26] Ger. Pat. 153,525; U.S. Pat. 849,900; Brit. Pat. 17,477/02.
[27] Ger. Pat. 168,977 (1902).

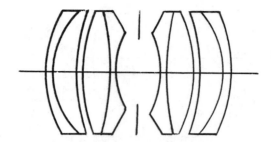

Figure 6.15. The Taylor-Hobson Series 15 lens.

Combination (d), in order (+ − − +), was used by Suter in his Double Anastigmat of 1902, the Berner Orthoskop of 1904, and the Sichel Plan-astigmat of 1904. System (e), in order (+ + − +), was used in the Goerz Pantar designed by F. Urban in 1904.[28]

E. Airspaced Quadruplets

Some lens designers felt that an improvement could be made by slightly separating the two halves of a quadruple double lens. An example was the Royal Anastigmat (probably designed by E. Gundlach), manufactured first by the Rochester Lens Company and later by Wollensak. Another example was the Taylor-Hobson Convertible Anastigmat Series 15[29] designed by H. W. Lee in 1931 (Fig. 6.15). One problem with this lens was a tendency to form strong ghost images of bright lights located in or close to the scene.

F. Quintuple Double Anastigmats

A few attempts have been made to design a five-element cemented component for a convertible double lens. The first of these was the Turner-Reich Anastigmat announced by the Gundlach Optical Company in 1895 (Fig. 6.16). Although this design was patented by H. H. Turner and J. C. Reich,[30] officers of the company, it was undoubtedly designed by Ernst Gundlach, as neither Turner nor Reich was a lens designer. If you compare the structure of the Zeiss Quadruple Protar Series VII (Fig. 6.14(a)) with the structure of the Turner-Reich lens, it will be seen that

---

[28] Ger. Pat. 171,369; U.S. Pat. 777,320; Brit. Pat. 11,108/04.
[29] Brit. Pat. 376,044 (1931).
[30] U.S. Pat. 539,370; Brit. Pat. 9,528/95.

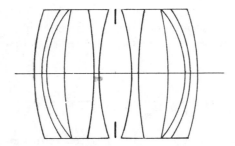

Figure 6.16. The Turner-Reich Anastigmat.

each consists of an old-achromat on the outside with a new-achromat doublet close to the stop. The difference between the two designs is that in the Zeiss lens the old-achromat is similar to half of a Rapid Rectilinear, while in the Turner-Reich lens the old-achromat resembles half of Gund-lach's Rapid Rectigraphic objective shown in Figure 4.14. As the perform-ances of the Rectigraphic and the Rectilinear were basically similar, we may expect to find that the Turner-Reich lens would be no better than the Zeiss Series VII, which is indeed the case. Actually, the rear component of the Zeiss objective had less coma than the rear of the Turner-Reich, so it might well have behaved better as a convertible unit. Throughout the history of photographic lenses it is common to find a designer patenting some novel lens that he could claim as his own, even though the design itself had no advantage over existing lenses. Also, many lenses are de-signed in a particular fashion for the sole purpose of avoiding the patents of some other company.

Another five-element lens was the Satz-Anastigmat of Goerz, designed by E. von Höegh in 1897 as a convertible lens (Fig. 6.17).[31] The high cost

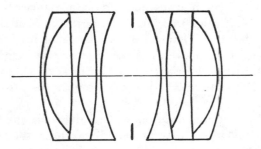

Figure 6.17. The Goerz Satz-Anastigmat.

---

[31] U.S. Pat. 599,700; Brit. Pat. 13,904/97.

of production, however, led to its early withdrawal. The problem of cementing five elements together with proper centering is horrendous, and it is likely that very few of these five-element cemented units were truly centered.

VI. THE DIALYTE TYPE

The old term *dialyte* refers to an airspaced achromatic doublet. When the elements in a cemented doublet are separated by a small finite distance, the powers of the lenses have to be increased to restore achromatism, the power of the negative element having to be increased more than the power of the positive element. This is exactly what is needed to reduce the Petzval sum of the lens, and von Höegh realized that this arrangement might lead to a useful anastigmat. As in his Dagor lens, he relied on symmetry to eliminate the three transverse aberrations, and he had five degrees of freedom in each half, namely, the powers of the elements, the shapes of the elements, and their separation. Using these degrees of freedom he was able to correct the Petzval sum, the chromatic aberration, and hold the focal length, while the shapes of the elements could be used to correct the spherical aberration and give a flat field. Assembling two of these units about a central stop completed the design.

The aberrations of a dialyte remain surprisingly constant over a wide range of object distances, and this type is often used for enlarging lenses. It also tends to be favorable for fairly high apertures but over a rather limited field. When used with a distant object, a little coma sometimes appears, which can be removed in the design by transferring some power from the front positive element to the rear positive element.

The original von Höegh lens was announced in 1899 as the Double Anastigmat Goerz Type B. It was made in two apertures: Series Ib at $f/4.5$ and Series Ic at $f/6.3$ (Fig. 6.18). In 1904 the names of all the Goerz lenses were changed; the type B lenses became the Celor for Series Ib and the Syntor for Series Ic. The lower-priced Syntor was manufactured until 1925, but in 1916 the Celor was replaced by the $f/4.5$ Dogmar designed by W. Zschokke (1870–1951).[32] An $f/6.3$ Dogmar was also manufactured for a short time.

In October 1901 Steinheil announced a dialyte lens similar to the Celor. This was called the Unofocal by virtue of the fact that the powers of all the

---

[32] Ger. Pat. 258,495; U.S. Pat. 1,108,307; Brit. Pat. 833/13.

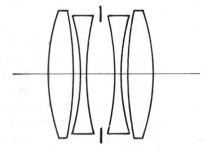

Figure 6.18. The Goerz Celor.

elements were approximately equal. This lens was much admired, and it was manufactured for many years by Steinheil and by Beck in England under license. The dialyte type of construction was adopted by Kodak in all their early anastigmats. It was also used in the Cooke Aviar lens de-signed by A. Warmisham,[33] which was used extensively during World War I for aerial photography.

In 1903 W. Zschokke of Goerz undertook the design of a symmetrical apochromatic process lens for the graphic arts. His first attempt, the Alethar, took the form of the basic dialyte type, but with cemented triplets in place of the inner negative elements (Fig. 6.19). The glasses used were dense barium crown and common crown, so it is doubtful if the maker's claim that this was an apochromat was justified. Furthermore, the actual glasses supplied by the manufacturer did not always come up to the catalog values, and the lens was soon discontinued. In the following year Zschokke designed the much simpler Artar, a basic dialyte design in which the

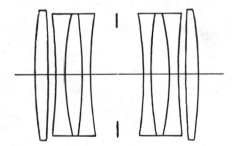

Figure 6.19. The Goerz Alethar.

---

[33] Brit. Pat. 113,590; U.S. Pat. 1,361,207 (1917).

positive elements were of dense barium crown and the negative elements were made of telescope flint glass having a favorable partial dispersion, so that the secondary spectrum could indeed be reduced. This lens remained the regular Goerz process lens for some seventy years.

In 1951 the C. P. Goerz American Optical Company announced the Huebner Artar Reverse Lens.[34] This consisted of a regular Apo Artar four-element lens, with a small Dove prism mounted in the middle in place of the iris diaphragm. This prism served to reverse the image left to right; further, the image could be rotated by merely turning the lens about its axis. This objective came in two sizes: a 24-inch at $f/11$ and a 30-inch at $f/12.5$.

VII. UNSYMMETRICAL DOUBLE ANASTIGMATS

Occasionally, for some reason, a designer will try the effect of combining two dissimilar cemented components about a central stop. It is hard to see the virtue of such an arrangement, except perhaps as an economy measure. Some examples of this are:

a. *Singlet front, triplet rear*
   Rodenstock: Imagonal, designed by Coblitz[35]
b. *Doublet front, triplet rear*
   Steinheil: Rapid Antiplanet[36]
   Turillon: Planigraphe of 1895
   Berner: Collar of 1904
c. *Doublet front, quadruplet rear*
   Rodenstock: Heligonal of 1905
   Staeble: Polyplast-Satz of 1929
d. *Triplet front, doublet rear*
   Leitz: Periplan[37]
   Ernemann: Anastigmat

---

[34] U.S. Pat. 2,408,855 (1942).
[35] Ger. Pat. 177,266 (1904).
[36] Ger. Pat. 76,662 (1893).
[37] Ger. Pat. 116,449; Brit. Pat. 3,041/99.

CHAPTER 7

The Triplet Lens and Its Modifications

I. THE COOKE TRIPLET

In 1893 H. Dennis Taylor (1862–1943), chief optical designer at Cooke of York, developed an entirely new type of triplet photographic objective. In a paper read before the Optical Society of London, Taylor explained the reasoning that led him to this new design.[1] His argument was that if we take a thin positive and a thin negative lens of equal power and place them in contact, the two lenses will neutralize each other, giving a system having zero power and zero Petzval sum. On separating the two lenses the system will acquire positive power but the Petzval sum will remain unchanged. Of course, such a highly unsymmetrical arrangement would possess horrendous oblique aberrations, so Taylor suggested splitting one of the elements into two and mounting one-half on each side of the other component. He patented both arrangements but he much preferred that in which the positive element was split and the negative element placed between the two halves. Actually, the other arrangement formed the basis of the recent

---

[1] H. D. Taylor, "Optical Designing as an Art." *Trans. Opt. Soc.* **24,** 143 (1923).

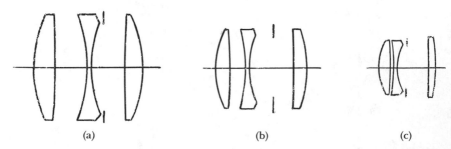

(a)   (b)   (c)

Figure 7.1. Three typical Cooke Triplets, drawn to the same scale: (a) $f/3$, ±22°; (b) $f/4.5$, ±24°; and (c) $f/5.6$, ±27°.

Zeiss Hologon lens in which the outer negative elements are deeply curved and of a thick meniscus shape (see Chapter 10, Section III).

At first Taylor thought that he would have to correct each of the three elements for spherical and chromatic aberrations, but he quickly found that this was unnecessary and that he had enough degrees of freedom to correct all aberrations with just three separated single elements. His first patent described an $f/4$ portrait lens with a field of only ±13° made from common crown and flint glasses (Fig. 7.1(a)), the two airspaces being approximately equal.[2] In 1895 he took out another patent in which he claimed a different design covering a wider field at a smaller aperture in which the front airspace was narrow and the rear airspace wide (Fig. 7.1(c)).[3] He also used high-index barium crown glasses and a lower-index flint in these later designs, one of which covered ±30° at $f/7.7$ and another ±26° at $f/5.6$. If the patent data can be trusted, these designs had excellent flat fields, but they showed signs of excessive astigmatism at intermediate field angles and a considerable amount of oblique spherical aberration, even though the axial image was well corrected in this regard.

To understand the logic of this design, the three lens powers and the two airspaces provide five degrees of freedom, which are used to control five quantities: the two chromatic aberrations, the Petzval sum, the focal length, and the desired ratio between the airspaces. This leaves the three lens shapes (bendings) to correct the spherical aberration, coma, and astigmatism. There is no symmetry to help the designer, and there is no control over distortion. If the distortion should turn out to be excessive, the only cure is to change the glasses and repeat the design.

[2] Brit. Pat. 22,607/93; U.S. Pat. 568,052; Ger. Pat. 81,825.
[3] Brit. Pat. 15,107/95; U.S. Pat. 568.052; Ger. Pat. 86,757.

Dennis Taylor worked entirely by algebraic formulae, which he developed himself, and he claimed that he never traced any rays.[4] When the design was as good as he could make it, the actual lens was fabricated and examination of the image on a lens-testing bench suggested changes that should be made to improve the performance. This is an expensive and slow process and Taylor was fortunate in having access to a good optical shop. It is hardly a procedure we would adopt today. In his patents Taylor also suggested variations of the simple three-element design, such as the use of a cemented doublet for the inner negative component or splitting it into two halves about a central stop.

Although the Triplet is essentially a simple system, it is difficult to manufacture because the elements are so strong, especially the inner negative element. Some of the early Cooke lenses were equipped with three centering screws by which the negative element could be centered by trial after the assembly was complete, the screws then being covered with cement to prevent further movement. Some manufacturers found that a four-element lens of the dialyte type was so much easier to assemble that they were reluctant to use Triplets at all. Nevertheless, these difficulties have been overcome and today Triplets of various kinds are used almost universally for all the lenses of intermediate aperture sold on modern cameras.

Because Taylor's own company, Cooke of York, did not wish to manufacture photographic lenses, Taylor took the design to a small optical and mechanical workshop that had been established a few years previously in Leicester called Taylor, Taylor and Hobson [no relation]. They gladly undertook to manufacture Triplet lenses, but they were to be known as Cooke Triplets out of respect for Taylor's employer.

The first Cooke Triplets sold by the Taylor-Hobson company fell into several groups:

Series	Aperture
II	$f/4.5$
III, VII	$f/6.5$
IV, VI	$f/5.6$
V	$f/8$ to 10

---

[4] H. D. Taylor, *A System of Applied Optics.* Macmillan, London, 1906.

The objectives of higher aperture were expensive items intended for portrait photography; some of the later versions were equipped with means for introducing a slight diffusion into the image by varying one of the airspaces. The Series V lenses were intended for copying and process work.

Since that time every manufacturer has made Triplet lenses under a wide variety of trade names and, surprisingly, there have been over eighty patents issued covering lenses of the three-element Cooke type. These patented designs differ from one another in the types of glass used and how the unavoidable aberration residuals are adjusted, but it is hard to regard them as "inventions." They appear to be routine designs that, in principle, could be generated automatically by a sufficiently complex computer program.

II. THE HELIAR AND ITS VARIATIONS

In 1900 Hans Harting of Voigtländer designed the original Heliar lens.[5] He was evidently trying to produce a symmetrical modification of the Cooke Triplet. If we divide the Triplet by an imaginary plane section through the middle of the inner negative element, we find that we do not have enough degrees of freedom in the rear half to correct all the longitudinal aberrations. So Harting replaced the rear single element of the Triplet by a cemented doublet, which then gave him a sufficient number of degrees of freedom to design the lens. In the rear component he had the

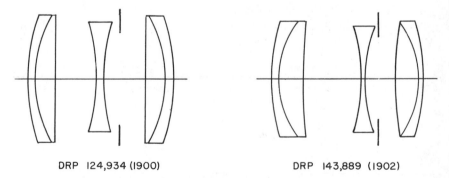

DRP 124,934 (1900)   DRP 143,889 (1902)

Figure 7.2. Two $f/4.5$ Heliar designs by Harting.

---

[5] Ger. Pat. 124,934; U.S. Pat. 716,035; Brit. Pat. 22,962/00.

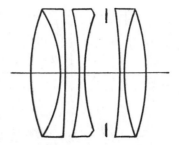

Figure 7.3. The Dynar.

concave surface of the divided negative element and there were three radii in the cemented positive component plus one airspace, giving five degrees of freedom with which to control the focal length and four longitudinal aberrations (spherical, chromatic, astigmatism, and the Petzval sum), leaving the eventual symmetry to correct the three transverse aberrations in the usual way. Any reasonable types of glass can be used in this design (Fig. 7.2(a)).

If the patent data are to be trusted, Harting's original design was not very good. It had a large Petzval sum and suffered from considerable astigmatism. The oblique spherical aberration typical of the Cooke Triplet was replaced by heavy coma when the lens was used with a distant object, and the overall performance must have been poor. Harting was evidently conscious of these shortcomings, for two years later he patented an unsymmetrical version of the lens (Fig. 7.2(b)).[6] The Petzval sum was greatly reduced and the astigmatism was well corrected, but the coma was as bad as before and the distortion was worse. By now the lens was about as good as a comparable Triplet, as indeed it should have been with five elements instead of the original three.

In 1903 Harting tried the effect of turning the outer components around so that the cemented interfaces were convex towards the stop instead of concave (Fig. 7.3).[7] He may have been influenced by the Tessar in which this arrangement of the rear component was used. The new lens was called the Dynar. The astigmatism was slightly worse than in the Heliar, but otherwise the new design was better. In the same year Harting patented a hybrid design called the Oxyn in which the front doublet was similar to that in the Heliar while the rear doublet resembled the Dynar

[6] Ger. Pat. 143,889; Brit. Pat. 13,441/02.
[7] Ger. Pat. 154,911; U.S. Pat. 765,006 (1903).

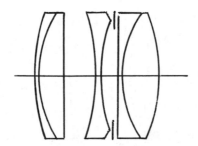

Figure 7.4. The Oxyn.

(Fig. 7.4).[8] This lens was sold as a copying objective with an aperture of $f/9$, at which low aperture the corrections were excellent.

From 1903 until after the first World War this type of construction was largely ignored by lens designers. In 1919 a Dynar-type objective was announced by Dallmeyer called the Pentac, designed by Lionel B. Booth.[9] It had an unusually high aperture of $f/2.9$. The field was slightly inward-curving, but in every other respect the lens was excellent. It is well known that a small amount of inward-curving field tends to reduce the astigmatism markedly, and this property has often been used by lens designers. The Pentac was much admired and was sold for many years.

After the first World War, Voigtländer revived the Dynar type of construction but, as they evidently felt that the name Heliar was preferable to Dynar, they called their new lenses Heliars and the name Dynar was abandoned. A few other designers have made us of this type, notably F. E. Altman of Kodak.[10] Five of his designs are illustrated in Figure 7.5; they were used in such applications as the lens for the Medalist camera and the Microfile Ektar.

III. THE SPEEDIC TYPE

One obvious way to increase the relative aperture of a Triplet lens is to split the strong rear positive element into two (Fig. 7.6). This procedure was tried in 1900 by Edward Bausch[11] and the same idea was popularized

---

[8] Ger. Pat. 154,910; U.S. Pat. 766,036 (1903).

[9] Brit. Pat. 151,506; U.S. Pat. 1,421,156 (1919).

[10] U.S. Pat. 2,279,384; Brit. Pat. 547,691 (1941).

[11] U.S. Pat. 660,747 (1900).

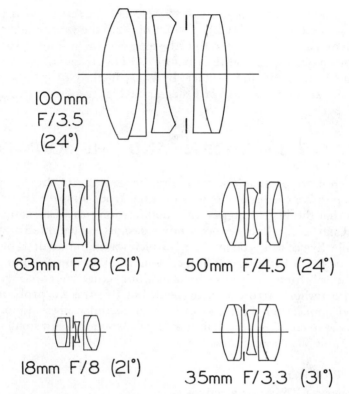

100mm
F/3.5
(24°)

63mm F/8 (21°)   50mm F/4.5 (24°)

18mm F/8 (21°)   35mm F/3.3 (31°)

Figure 7.5. Five Heliar-type lens designs by Altman of Kodak.

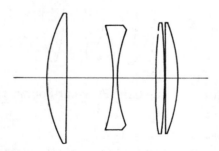

Figure 7.6. The Speedic lens.

by H. W. Lee[12] in his Speedic lens of 1924. The process of the design has been described by Lee.[13] At an aperture of $f/2.5$ the spherical aberration of this lens was well corrected, but the astigmatism at intermediate field zones was more noticeable than before. Although several lenses of this type have been made, for instance, by W. F. Bielicke of the Astro Company, there are better ways to make lenses having an aperture greater than about $f/3$.

IV. THE ERNOSTARS AND THE SONNARS

A much more profitable way to raise the aperture of the Triplet is to insert a positive meniscus element into the front airspace (Fig. 7.7). It appears that the first to suggest this modification was a Chicago optician named Charles C. Minor. Among other designs, he patented a four-element objective of this type in 1916,[14] which was manufactured by Gundlach as the $f/1.9$ Ultrastigmat in focal lengths of 40 mm, 50 mm, and 75 mm, primarily for use on professional motion-picture cameras (Fig. 7.8). It covered a rather narrow angular field, but this was no problem in its intended application. Later, several other manufacturers adopted this simple type of construction, mainly as an $f/1.9$ objective for small motion-picture cameras. Among these were:

Agfa: Prolinear
Angénieux: All Y-type objectives
Bell & Howell: Lumax

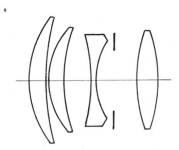

Figure 7.7. The basic Ernostar-Sonnar type.

---

[12] Brit. Pat. 224,425; U.S. Pat. 1,739,512 (1924).
[13] H. W. Lee, "The Taylor-Hobson $f/2.5$ Anastigmat Lens." *Proc. Opt. Conv. 1926,* page 851.
[14] U.S. Pat. 1,360,667; Brit. Pat. 187,082 (1916).

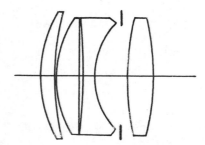

Figure 7.8. The Gundlach Ultrastigmat.

Enna: Kinaston
Isco: Kiptagon
Kodak: Cine Ektar
Meyer: Optimat, Primagon
Rodenstock: Lumar, Ronar
Taylor-Hobson: Ivotal, Serital, Super Comat
Wray: Unilux
Zeiss: Sonnar

In 1919 Ludwig Bertele (1900–1985) of the Ernemann Company started work on this type of construction. He eventually decided to use cemented doublets for the front two positive elements and in this way, although he was only 23 at the time and practically self-taught, he developed the epoch-making $f/2$ Ernostar lens, which was fitted to the small Ermanox camera and placed on the market (Fig. 7.9(a)).[15] This was the first camera to have sufficient speed and image quality for candid photography by available light. The pictures of prominent political figures taken with it by Erich Salomon are famous.

In the following year, Bertele succeeded in raising the aperture to $f/1.8$ with a slight widening of the field in his improved Ernostar design.[16] At the same time he developed a simpler model for use at $f/2.7$.[17] The three designs are shown together in Figure 7.9.

In 1926 the Ernemann Company was taken over by Zeiss-Ikon, and

---

[15] Ger. Pat. 401,274; U.S. Pat. 1,584,271; Brit. Pat. 191,702 (1923).
[16] Ger. Pat. 436,260; U.S. Pat. 1,708,863; Brit. Pat. 237,529 (1924).
[17] Brit. Pat. 237,861 (1924).

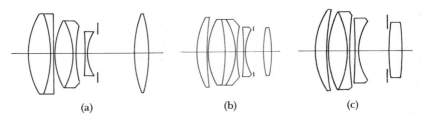

Figure 7.9. Three Ernostar designs: (a) the $f/2$ of 1923; (b) the $f/1.8$ of 1924; and (c) the $f/2.7$ of 1924.

Bertele became a Zeiss designer. The Ernostar lenses were continued in production under their old names for a few years and then dropped. In 1930 Bertele began the development of a series of excellent lenses called the Sonnars, based generally on the second Ernostar type (i.e., each lens had a single positive element in front followed by a thick negative menis-cus-shaped component, with a positive element behind). In the high-aperture Sonnars the thick negative component consisted of a cemented triplet, with high-index elements outside and a lower-index element between them (Fig. 7.10). In this respect the Sonnars resembled the Ultrastigmat, except that the narrow space between the second and third elements was filled with low-index glass instead of air. It has been mentioned previously that a cemented interface can be replaced by a pair of much weaker glass-air surfaces with reduced aberrations. In the Sonnar, however, the designer needed a strong cemented interface to control the higher-order aberrations, so he deliberately reversed the usual process and replaced air by glass.

In 1931 Bertele patented an $f/2$ Sonnar,[18] which was followed in the next year by an $f/1.5$ version.[19] The latter lens contained a strong cemented interface in the rear component (Fig. 7.10(b)) that separated two glasses having a very small difference in refractive index. This device had a powerful influence on the higher-order spherical aberration — exactly what was needed in a lens of this high relative aperture. The Zeiss Company has continued to manufacture Sonnar-type lenses, some of which are of lower aperture and longer focal lengths. A few other companies have used this type of construction but most have preferred the double-Gauss type for their high-aperture objectives.

It is interesting to note that the name Sonnar had been used previously

---

[18] Ger. Pat. 570,983; U.S. Pat. 1,998,704; Brit. Pat. 383,591 (1931).
[19] Ger. Pat. 673,861; U.S. Pat. 1,975,678 (1932).

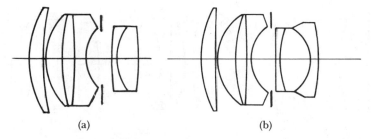

Figure 7.10. The original Sonnar designs: (a) $f/2$ and (b) $f/1.5$.

by the Contessa Company for one of their folding cameras and for the Tessar-type lens fitted to it. After Contessa became part of Zeiss-Ikon the name Sonnar became Zeiss's property.

V. MODIFIED TRIPLETS BY LEITZ

Most of the early lenses made for the Leica camera were modified Triplets with cemented components used in place of the single elements, mainly designed by Max Berek. The five principal classes of these modified Triplets are shown in Figure 7.11. They are:

(a) *1-C-1: Cemented inner negative component*
　　Hektor　　135 mm $f/4.5$ (1934)
　　Thambar　90 mm $f/2.2$ (1935)
　　Hektor　　125 mm $f/2.5$ (1954)
(b) *1-1-C: Cemented rear component* (Tessar type)
　　Elmar　　　50 mm $f/3.5$ (1925)[20]
　　Elmar　　135 mm $f/4.5$ (1931)
　　Elmar　　　35 mm $f/3.5$ (1931)
　　Elmar　　　90 mm $f/4.0$ (1932)
　　Elmar　　105 mm $f/6.3$ (1932)
　　Elmar　　　65 mm $f/3.5$ (1959)
　　Elmar　　　50 mm $f/2.8$ (1959)
(c) *1-C-C: Cemented middle and rear components*
　　Elmarit　　90 mm $f/2.8$ (1958)[21]

---

[20] Ger. Pat. 343,086 (1920). In Cox, *Photographic Optics*, 15th ed., some 68 sections of Triplets and related types have been pictured and listed.
[21] U.S. Pat. 2,995,980; Brit. Pat. 839,723 (1956).

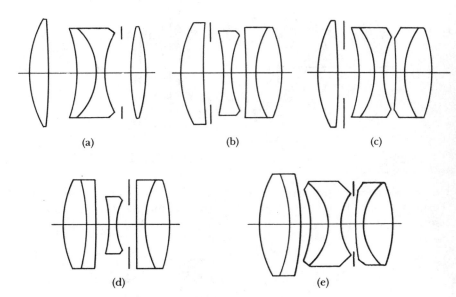

(a)   (b)   (c)

(d)   (e)

Figure 7.11. Five early lenses for the Leica camera.

(d) *C-1-C: Cemented front and rear components* (Heliar type)
   Hektor   28 mm $f/6.3$ (1935)
(e) *C-C-C: All three components cemented*
   Hektor   50 mm $f/2.5$ (1931)[22]
   Hektor   73 mm $f/1.9$ (1932)[23]

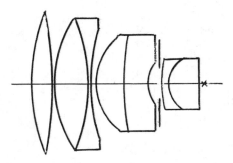

Figure 7.12. The Wray 16:1 radiographic copying lens, aperture $f/0.71$.

---

[22] Ger. Pat. 526,307 and Ger. Pat. 526,308 (1930).
[23] Ger. Pat. 585,456; U.S. Pat. 1,939,098; Brit. Pat. 382,246 (1931).

VI. THE WRAY HIGH-APERTURE RADIOGRAPHIC LENS

In 1950 C. G. Wynne of the Wray Company in England designed a very high aperture copying lens for use primarily in reproducing X ray images at a demagnification of 16:1.[24] It is shown in Figure 7.12. The front portion was similar to the 2-inch $f/1$ lens designed by Wynne for copying a CRT image at 4:1, but the rear portion consisted of a thick cemented doublet located very close to the image. This lens was also available in other focal lengths for use with an infinity object distance.

---

[24] Brit. Pat. 696,902 (1950); see also *J. Sci. Inst.* **28,** 172 (1951).

CHAPTER 8

Meniscus Anastigmats

I. THE DOUBLE-GAUSS LENS

Almost every high-aperture objective used today is of the type known familiarly as the Double-Gauss lens.[1] C. F. Gauss (1777–1855) was a famous German mathematician and actually had little to do with this lens. As early as 1817, Gauss had described a novel type of telescope objective consisting of a pair of meniscus-shaped elements, one positive and one negative (Fig. 8.1). The virtue of this arrangement was that the spherical aberration was insensitive to the wavelength of the light, but for various reasons this type has seldom been used in astronomy. Indeed, the Alvan Clarks of Cambridge, Massachusetts, tried it only once, in a 9½-inch lens for Princeton made in 1877, but they felt that the result did not justify the difficulties of manufacture.[2] However, in 1888, Alvan G. Clark, the son of

---

[1] A. Cox, *Photographic Optics*, 15th ed., page 485. Amphoto, Garden City, 1974. Sectional diagrams of nearly 70 lenses of the double-Gauss and related lenses are shown here. See also K. Yamada, "Japanese Photographic Objectives," *Phot. Sci. and Eng.* **2**, 6 (1958).

[2] D. J. Warner, *Alvan Clark and Sons, Artists in Optics*, pages 28 and 85. Smithsonian, Washington, D.C., 1968.

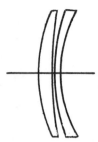

Figure 8.1. The original Gauss telescope objective.

the founder, must have thought that there was some good in the scheme, for he took out a patent in which he suggested that a pair of Gauss objectives mounted back-to-back in a tube might make a possible photographic objective.[3] As he only suggested some radii of curvature, with no statement as to the thicknesses or separations or the types of glass to be used, it is difficult to judge whether his idea had any possibility of success. Apparently Bausch & Lomb investigated the matter and in their catalogs from 1890 to 1898 they included an Alvan G. Clark lens (Fig. 8.2), with apertures of $f/8$, $f/12$, and $f/35$, which they offered for sale as wide-angle objectives. These were not successful, however, and they were dropped from the catalog, but the idea was taken up by various European manufacturers under such names as the Omnar of Busch, the Aristostigmat by Meyer, and the Ross Homocentric. Some of these lenses remained in the catalogs until 1930; they were made in a range of apertures up to $f/4$.

An extreme example of the simple double-Gauss type was the Zeiss

Figure 8.2. Alvan G. Clark lens, from Bausch & Lomb's catalog.

---

[3] U.S. Pat. 399,499 (1888).

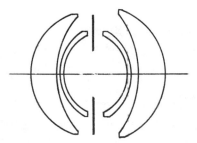

Figure 8.3. The Metrogon lens.

Topogon, which was capable of covering a full 90° field at $f/6.3$. It was designed by Robert Richter in 1933[4] and was used for many years as the standard aerial camera lens, particularly as the distortion was small, though it was not negligible. Thousands of these lenses were made during and after World War II, both in Germany in its original form and in the United States as the slightly modified Bausch & Lomb Metrogon (Fig. 8.3). In the 6-inch size, this covered a 9″ × 9″ format, but the vignetting was so great that it was necessary to mount a graded-density filter in front of the lens to hold back the central illumination while permitting the whole of the light to reach the edges of the picture area. A few Metrogon lenses were also made in the 12-inch size to cover an 18″ × 18″ format. Similar designs were made by other manufacturers in Europe.

Here is a partial listing of lenses consisting of four single meniscus elements:

Bausch & Lomb: Metrogon, Process Anastigmat
Boyer: Perle
Busch: Omnar
Dallmeyer: Wide-Angle Anastigmat
Goerz: Geotar, Rectagon
Hermagis: Dellor Series D
Ilex: Anastigmat Series D
Kodak: Wide-field Ektar
Laack: Wide-angle Dialytar
Meyer: Aristostigmat
Plaubel: Wide-angle Orthar, Pecostigmat

---

[4] Ger. Pat. 636,167; U.S. Pat. 2,031,792; Brit. Pat. 423,156 (1933).

Rietzschel: Dialyte
Rodenstock: Eurynar, Lumar, Ronar
Ross: Homocentric
Schneider: Isconar
SOM: Aquilor
Wollensak: Wide-angle Raptar
Wray: Wide-angle Copying Lens
Zeiss: Hekla, Topogon

An interesting development of this type has been the insertion of a thin negative element into the central airspace of a double Gauss lens (Fig. 8.4). This arrangement was pioneered by Beck of London in their Isostigmar lenses announced in 1906.[5] Seven series of Isostigmars were manufactured, as follows:

Series I	$f/4.5$	±30°	3 to 12 inches focal length.
Ia	6.5	30°	9 to 19
II	5.8	35°	3 to 8
III	7.7	32°	5 to 9
IV	6.3	45°	3 to 7
V	11.0	30°	12 to 30
VI	5.6	30°	10 to 17

The stated angular fields were probably optimistic and would be attainable only at small stops. The makers claimed that each half of these lenses could be used independently at a lower aperture.

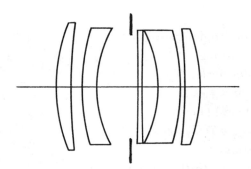

Figure 8.4. The Beck Isostigmar.

---

[5] Sectional diagrams of these lenses are given in *Brit. J. Almanac,* 1911, pages 108–112. See also Brit. Pats. 27,180/06, 14,673/08; U.S. Pat. 871,559; Ger. Pat. 194,267.

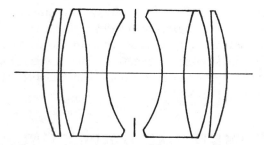

Figure 8.5. The Planar, by P. Rudolph.

II. THE ZEISS PLANAR LENS

Finding that his $f/4.5$ Anastigmat lens was not fully satisfactory, in 1896 Paul Rudolph turned his attention to the double-Gauss type. This lens had the advantage of symmetry, thus saving the designer a great deal of work in trying to correct the three transverse aberrations. However, as ordinarily constructed, the wide airspace separating the thin positive and negative elements in the double-Gauss design led to a large amount of oblique spherical aberration and a noticeable "belly" between the sagittal and tangential astigmatic images at intermediate field angles. Even at $f/8$ these defects were noticeable, and of course the oblique spherical aberration would become much worse at $f/4.5$. For some reason, it occurred to Rudolph to try thickening the negative elements and reducing the airspace between positive and negative elements as much as possible. The result was magical, as it had the desired effect of reducing both these defects simultaneously. However, when he came to achromatize the system, he found that there were no flint glasses available with a sufficiently high dispersive power.

Rudolph, therefore, had the ingenious idea of inserting a "buried surface" into the thick negative elements. A buried surface is a cemented interface separating two types of glass having the same refractive index but quite different dispersive powers. By varying the radius of curvature of the buried surface, almost any amount of chromatic aberration could be introduced at will with no effect whatever on the other aberrations.[6] In this design, Rudolph used two glasses having a refractive index of 1.57, with which he produced an excellent $f/4.5$ objective called the Planar (Fig. 8.5)

---

[6] Ger. Pat. 92,313; U.S. Pat. 583,336; Brit. Pat. 27,635/96.

that was manufactured by Zeiss for many years. Later, an Apo-Planar was introduced for color photography in the graphic arts.

There is some evidence that G. Richmond, in his Ross Homocentric design, also tried thickening the negative elements, as Rudolph had done, but at $f/6.3$ he did not need to use a buried surface for achromatism. His selection of glasses is not known.

III. UNSYMMETRICAL DOUBLE-GAUSS LENSES

After this initial flurry, the double-Gauss type was ignored by lens designers until 1920 when H. W. Lee of the Taylor-Hobson Company succeeded in raising the aperture to $f/2$ with a field coverage of $\pm 23°$ for use on ordinary cameras. The procedure he followed in designing this lens has been described.[7] Lee's lens differed from the Planar in being unsymmetrical and by the use of crown glasses having a higher index than that of the flints (Fig. 8.6). It was known as the Opic or Series 0 lens and was included in Taylor-Hobson's catalogs for many years. However, the Ernostar lens that appeared three years later was much more successful as an $f/2$ camera lens, probably because the Opic was offered as just another lens while the Ernostar was sold in a complete camera with a focal-plane shutter and an excellent focusing means.

Following Lee's lead, other designers began to realize the virtues of this type of construction. In 1925 A. Tronnier designed the Schneider $f/2$

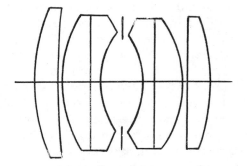

Figure 8.6. The Lee Opic lens.

[7] H. W. Lee, "The Taylor-Hobson $f/2$ Anastigmat," *Trans. Opt. Soc.* **25,** 240 (1924); also, Brit. Pat. 157,040 (1920).

[8] Ger. Pat. 439,556 (1925).

Xenon lens,[8] and two years later W. Merté of Zeiss designed a series of Biotar lenses, including a 50 mm $f/2$ lens for 35 mm cameras and a 25 mm $f/1.4$ for 16 mm movies. In 1931 Lee designed the very successful $f/2$ Speed Panchro,[9] for many years the standard lens used by Hollywood cameramen. In 1933 M. Berek designed the $f/2$ Summar lens for the Leica camera.[10] The design of the basic double-Gauss lens has been discussed by Mandler.[11]

A partial listing of lenses of the classical six-element double-Gauss type follows. However, as always in such lists, the reader must understand that the same trade name is often used for lenses of quite different construction, and a name given in this list may apply to a variety of construction types and not solely to double-Gauss lenses:

Agfa: Soligon
Angénieux: S-type
Astro: Kino, Tachar
Bausch & Lomb: Animar, Baltar, Raytar
Boyer: Saphir
Dallmeyer: Super Six
Enna: Annaston
Isco: Westagon
Kinoptik: Apochromat, Fulgior
Kodak: Ektar
Leitz: Dygon, Summar
Meyer: Domiron
Rodenstock: Heligon
Ross: Xtralux
Schneider: Xenon, Xenogon
Steinheil: Quinon
Taylor-Hobson: Amotal, Ivotal, Kinic, Opic, Panchrotal, Speed Panchro
Wollensak: Raptar
Wray: Copying lens
Zeiss: Biotar, Flexon, Planar

---

[9] Brit. Pat. 377,537; U.S. Pat. 1,955,591 (1931).
[10] Brit. Pat. 423,468 (1933).
[11] W. Mandler, *Proc. SPIE.* **237**, 222 (1980).

In the early 1930s, many variations of the basic Planar type began to appear. Most of these involved separating the elements of the front negative component and bending them independently. In several high-aperture lenses, the rear positive element has been split into two, with each element bent independently. Sometimes one or both of the outer positive elements has been replaced by a cemented doublet, while in some instances two or more of these variations have been used simultaneously.

In Figure 8.8 is shown a group of lenses sold by Leitz between 1938 and 1968 for use on various models of the Leica camera. The 50 mm $f/2$ Summar was the first (Fig. 8.8(a)), but users complained of the poor definition at full aperture, so it was replaced by the seven-element Summitar in 1939 (Fig. 8.8(e)).[12] This lens was similar to the Kodak $f/1.9$ Ektar used on the Ektra camera.[13] This group of designs well illustrates some of the attempts that have been made to improve the basic double-Gauss form for high-aperture lenses.

During World War II, G. Aklin at Kodak designed a seven-element version of the Gauss type known as the $f/2.5$ Aero Ektar (Fig. 8.7).[14] This was one of the first lenses to make use of the recently developed lanthanum crown glass. It was made in large numbers in the 7-inch size to cover a 4¼″ × 4¼″ format, and a few lenses were made in the 12-inch size for a 9″ × 9″ image.

The number of double-Gauss lenses of various forms that has been manufactured is legion. Over 300 patents for lenses of this type are known, and certainly many lenses that have been manufactured were never pat-

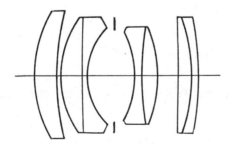

Figure 8.7. The Kodak 7-inch and 12-inch Aero Ektar lens.

[12] Ger. Pat. 685,572 (1936).
[13] U.S. Pat. 2,262,985; Brit. Pat. 548,252 (1940).
[14] U.S. Pat. 2,343,627; Brit. Pats. 559,379 and 559,390 (1941).

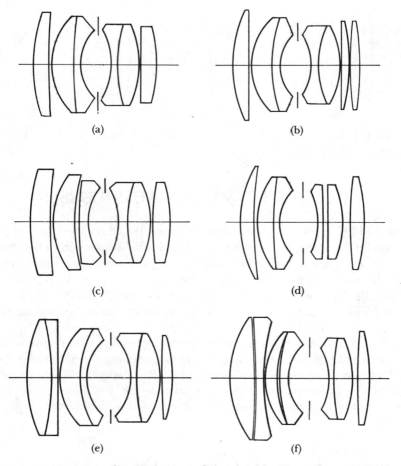

Figure 8.8. A group of Leitz lenses of the double-Gauss form:
(a) Summar $f/2$ (1933);   (b) Summarit $f/1.5$ (1949);
(c) Summicron 90 mm $f/2$ (1958);   (d) Summicron $f/2$ (1968);
(e) Summitar $f/2$ (1939); and   (f) Summicron $f/2$ (1953).

ented. Today almost all of the high-aperture lenses supplied on Japanese cameras are of this type or one of its modifications, with apertures as high as $f/0.95$ in some cases. An interesting example is the $f/1.2$ Leitz Noktilux,[15] which had the usual six elements but whose two outer surfaces were aspheric.

---

[15] U.S. Pat. 3,459,468; Brit. Pat. 1,088,192 (1964).

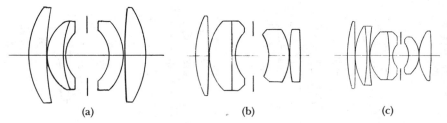

Figure 8.9. Three Wynne designs: (a) the $f/2$ Unilite; (b) the $f/1.9$ Cine Unilite; and (c) the $f/1$ CRT lens.

IV. FIVE-ELEMENT GAUSS LENSES

In 1944 C. G. Wynne, soon after he joined Wray, realized that he could eliminate the rear positive element of the double-Gauss design by separating the rear negative doublet and bending the negative element of the doublet into a deep meniscus form concave to the stop. Using this technique, he designed three outstanding lenses, the $f/2$ Unilite,[16] the $f/1.9$ Cine Unilite,[17] and the $f/1$ CRT lens for use at a magnification of $4:1$.[18] These lenses are illustrated in Figure 8.9.

After World War II, the same idea was adopted by other manufacturers for lenses of $f/2.8$ aperture, a typical example being the Schneider Xenotar[19] (Fig. 8.10) and the Zeiss series of so-called Planar lenses. Some of the Planars had the two single elements in the rear and some in the front. Examples of U.S. patents displaying lenses of this five-element type are:

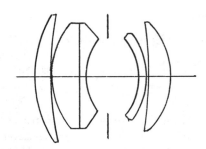

Figure 8.10. The Schneider Xenotar.

---

[16] Brit. Pat. 575,076; U.S. Pat. 2,499,264 (1944).
[17] Brit. Pat. 575,075; U.S. Pat. 2,487,749 (1944).
[18] Brit. Pat. 604,883; U.S. Pat. 2,487,750 (1945).
[19] U.S. Pat. 2,683,398 (1952).

(a) Single elements in the rear:

    Tronnier 2,720,139 (1953)
    Lange   2,724,994 (1953)
    Klemt   2,831,395 (1954)
    Zollner   2,968,221 (1959)

(b) Single elements in the front:

    Berger   2,744,447 (1953)
    Lange   2,799,207 (1954)

Another well-known Zeiss lens of this general type is the Biometar.

V. HYBRID TYPES

A few lenses have been designed in which the Gauss type has been used
for one half and another type for the other half. Rudolph's Unar lens was
an example of this arrangement (see Fig. 6.5). In this lens, the rear half was
of the simple Gauss type involving two single meniscus elements, while the
front half was a dialyte with a biconvex crown and a biconcave flint. The
$f/2.7$ Miniature Plasmat[20] designed by Rudolph in 1931 was another
example in which the front component was of the ordinary Gauss type
while the rear component was the rear half of a Plasmat (Fig. 8.11). This
lens had no particular virtues and it was manufactured for only a very short
time. A much better design was the earlier $f/2.9$ Makro Plasmat of 1926,
in which the three elements in the rear half were completely airspaced (Fig.
8.12).[21] This lens was made for many years and proved to be surprisingly

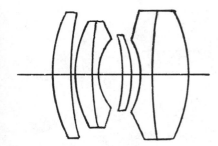
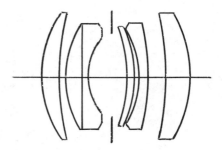

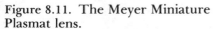

Figure 8.11. The Meyer Miniature
Plasmat lens.

Figure 8.12. The Makro Plasmat.

---

[20] Ger. Pat. 572,222; U.S. Pat. 1,945,570 (1931).
[21] Ger. Pat. 456,912; U.S. Pat. 1,812,717; Brit. Pat. 261,326 (1926).

popular. The prefix Makro suggests that its performance held up well for close object distances.

It will be noticed that the Miniature Plasmat shown in Fig. 8.11 contained four components in order $(+-:+-)$ with the stop in the middle. This lens arrangement was revived by Bertele in his design for a 35mm $f/2.7$ lens to be used on the Contax camera.[22] This lens was called the Biogon, a name that was used later by Zeiss for a very different type of construction (see page 150). Actually, this first Biogon was a modification of Bertele's $f/1.5$ Sonnar, shown in Figure 7.10(b) on page 113. The inside triplet was reduced in size while the rear triplet was greatly enlarged and the three elements airspaced (Fig. 8.13). Bertele evidently liked this arrangement, as he modified it somewhat when designing his 1944 Aviotar lens, manufactured by Wild for aerial photogammetry.[23] It was an excellent design, covering $\pm 30°$ at $f/4.2$ with virtually no distortion.

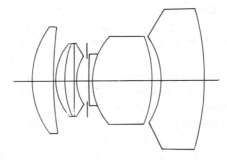

Figure 8.13. The first Zeiss Biogon.

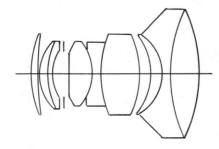

Figure 8.14. The Wild Aviotar lens.

---

[22] Ger. Pat. 652,062; U.S. Pat. 2,084,309; Brit. Pat. 459,739 (1934).
[23] U.S. Pat. 2,549,159; Brit. Pat. 660,259 (1947).

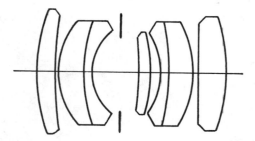

Figure 8.15. The Leitz Elcan.

A few designers have experimented with inserting a thin positive menis-cus element into the center space of a 6-element double Gauss lens. Exam-ples of this are the Leitz Elcan objective (Fig. 8.15) designed by Mandler in 1958, and the similar 35mm $f/1.4$ Leitz Summilux of 1960 (Fig. 8.16).[24] Alternatively, one may regard these two designs as a Miniature Plasmat with the addition of a positive element at the rear. Without some knowl-edge of the procedure adopted by the designer, it is sometimes difficult to assign a lens to some particular structural type.

It will be found that many modern lenses do not fit into any recognized class; these are obviously the result of computer optimization, as the com-puter generally has no knowledge of the classical lens types.

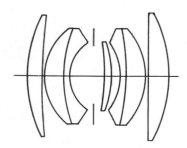

Figure 8.16. The Leitz 35 mm $f/1.4$ Summilux.

---

[24] U.S. Pat. 2,975,673; Brit. Pat. 867,266 (1958).

CHAPTER 9

Telephoto Lenses

The idea of inserting a negative lens between an ordinary objective and the focal plane to magnify the image is old; indeed, it is said to have been used by Kepler.[1] Early in the last century, Peter Barlow (1776–1862), the author of the well-known Barlow's Tables, developed a negative achromat for use in small telescopes to enlarge the image. Barlow lenses are still being sold and used extensively today. The combination of a positive front component and a negative rear component constitutes a telephoto lens. Many papers have been written on the telephoto lens, which has enjoyed an interesting history.[2-5]

The shortness of a telephoto as compared with an ordinary photographic objective is indicated in Figure 9.1. In (a) an $f/8$ Triplet is

---

[1] J. Waterhouse, "The Early History of Telephotography," *Proc. Opt. Conv. of 1905,* page 115. Norgate and Williams, London, 1905.

[2] L. B. Booth, "The Telephoto Lens." *Proc. Opt. Conv. 1926,* page 861.

[3] H. W. Lee, "The Development of the Telephotographic Objective." *Ibid,* page 869.

[4] C. F. Lan-Davis, *Telephotography.* Pitman, London, 1947.

[5] R. Kingslake, "Telephoto vs Ordinary Lenses." *JSMPTE* **75,** 1165 (1966). In Cox, *Photographic Optics,* 15th ed., pages 534ff, sectional diagrams of 58 different telephoto lenses are shown.

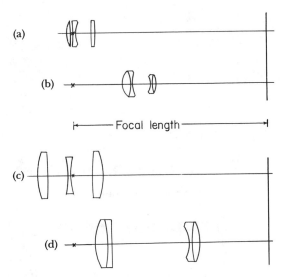

Figure 9.1. Illustrates the shortness of a telephoto lens: (a) and (b) $f/8$; (c) and (d) $f/3.8$.

compared with an $f/8$ telephoto in (b). In (c) is shown an $f/3.8$ Triplet alongside an $f/3.8$ telephoto in (d). In each diagram the "x" indicates the position of the rear nodal point from which the focal length is measured.

At first, the effectiveness of a telephoto lens was expressed by the ratio of the focal length to the back focal distance, because this was a measure of the reduction of bellows length of the camera. However, this was really a meaningless quantity, as, for example, the ordinary Petzval lens had a short back focus but it was by no means a telephoto. A much more meaningful expression is the ratio of the total length from the front lens vertex to the focal plane to the focal length. This ratio is now universally employed; it is known as the telephoto ratio of the lens. Its value is commonly about 80 percent, but in some cases it has been reduced to almost 60 percent by a great increase in the powers of the two components, with, of course, a corresponding increase in the various residual aberrations.

I. THE PETZVAL ORTHOSKOP

When Petzval was working on his famous Portrait lens, he designed at the same time a landscape lens covering a wider field at a smaller aperture. He used the same front component for both lenses, but the rear component of the landscape lens was negative in power (see Fig. 3.2). As the rear nodal point was slightly ahead of the front lens surface, this system would now be called a telephoto, but that name had not yet been introduced. Petzval submitted both designs to the optician P. W. F. Voigtländer (1812–1878), with all technical details but no binding agreement. The Portrait lens was manufactured at once, but no work was done on the landscape lens at that time. Later, when photography was well established, the need for a distortionless landscape lens became acute, and in 1854 Petzval, having quarreled with Voigtländer, commissioned the Viennese optician Dietzler to manufacture it for him. This was done, and the new lens was announced in 1856 as the Photographic Dialyte. It immediately became popular with photographers, who regarded it as the best distortionless landscape then available.

As soon as the new lens appeared, Voigtländer recognized it as the second design that Petzval had given him in 1840 and which he had forgotten. So he resurrected the lens data, and put the lens on the market under the name of Orthoskop, given in recognition of its low distortion.[6] The new lens was sold extensively, and a version of it was made by Harrison in New York. Indeed, it became so famous that Dietzler had to change the name of his lens from Dialyte to Orthoskop to compete with Voigtländer! Naturally, a quarrel arose (not by any means the first) between Petzval and Voigtländer, but in the absence of any legal agreement there was little either could do about it.

At first, photographers were delighted with the Orthoskop, but soon dissatisfaction arose. Careful tests showed that the new lens was not absolutely free from distortion and the field was noticeably backward-curving, so the lens was forgotten as quickly as it had appeared.

II. THE DALLMEYER AND MIETHE TELEPHOTOS

In 1891 one of those simultaneous and independent inventions that have appeared several times in the history of camera lenses occurred. In October 1891, T. R. Dallmeyer (1859–1906), the son of J. H. Dallmeyer, applied for a British patent on a telephoto lens that he had designed.[7] It consisted of a cemented doublet in front and a cemented triplet behind,

---

[6] M. von Rohr, *Theorie und Geschichte des photographischen Objektivs*, page 250. Springer, Berlin, 1899.

[7] Brit. Pat. 16,780/91 (not completed).

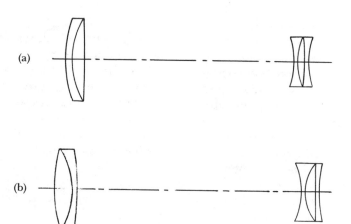

Figure 9.2. Two telephoto lenses from 1891: (a) Dallmeyer and (b) Miethe.

the two components being mounted with a variable separation so that the focal length could be changed in a continuous manner (Fig. 9.2(a)). However, this cannot be regarded as a zoom lens because the bellows extension of the camera had to be altered to suit the particular lens separation that was being used. The relative aperture also varied with the lens extension. Such a variable-focal-length system was often referred to as Pancratic.

At almost exactly the same time, Adolph Miethe (1862–1927) applied for a German patent on a virtually identical arrangement (Fig. 9.2(b)).[8] In successive issues of the *British Journal of Photography*, from October to December 1891, the two rival inventors addressed letters to each other, trying to sort out who was the original inventor. They could not reach an agreement and neither patent was ever issued.

Later, in 1891, Dallmeyer applied for a British patent[9] on an improved design (Fig. 9.3) in which he used an ordinary portrait lens in front and a pair of negative cemented doublets behind, the separation being adjustable as before. Two rear components of different powers were made available and the system was sold for many years. At that time enlargers were not generally used because of the lack of a convenient light source, and photographers were quick to appreciate the advantages of having a long focal length available without the need for a long bellows extension. Many manufacturers began to offer telenegative attachments for use be-

---

[8] J. M. Eder, *Die photographischen Objektive*, page 159. Knapp, Halle, 1911.
[9] Brit. Pat. 21,933/91.

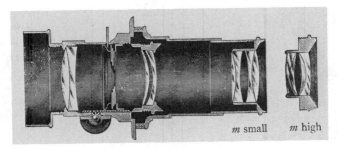

Figure 9.3. The Dallmeyer telephoto lens.

hind a normal camera lens (see Chapter 13, Section IIIB). For some reason, Dallmeyer felt that there was something magical about the telephoto construction and he wrote an entire book about it.[10] Of course, all lenses of the same focal length produce an image of the same size, but it was the variable focal-length feature that made the telephoto so attractive to photographers.

Varifocal projection lenses of the telephoto type have long been used on slide projectors, the advantage being that the projector can be set up at a convenient distance from the screen and the picture size then adjusted to fill the screen. A lens for this purpose was patented by Knipe in 1891[11] and a similar system was manufactured for many years by Beck,[12] known as the Multifex. The pincushion distortion of a telephoto lens, considered inevitable at the time, was converted into barrel distortion on the screen.

III. SELF-CONTAINED TELEPHOTOS

The real problem with these early telephotos was that the front component was a well-corrected regular photographic objective, and consequently the negative rear component should have been equally well corrected for all aberrations, which it certainly was not. Furthermore, the rear component has the effect of magnifying the residual errors of the front component and also lowering its aperture, so the front component has to have a high relative aperture and be exceptionally well corrected.

To avoid this difficulty, some manufacturers began to develop self-con-

---

[10] T. R. Dallmeyer, *Telephotography*. Heinemann, London, 1899.
[11] Ger. Pat. 88,889; U.S. Pat. 576,896; Brit. Pat. 12,219/96.
[12] Brit. Pat. 15,732/05.

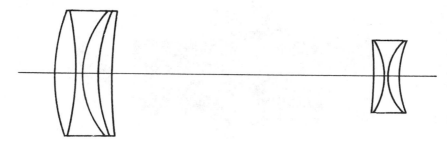

Figure 9.4. The Zeiss Tele-Tubus.

tained telephoto lenses that were designed as a unit at the most popular separation, hoping that the aberrations would not vary too much at other separations. Typical of these designs was the Zeiss Tele-Tubus, designed in 1896 and shown in Figure 9.4. The front component was an $f/3$ cemented quadruplet designed by Paul Rudolph,[13] and the rear was a cemented negative triplet. In their 1901 catalog, Zeiss listed three focal lengths of the front component and six focal lengths of the rear, giving a large number of possible combinations. In addition, it was possible to combine the various rear components with a Protar or other normal photographic objective, if desired. In their 1910 catalogue, the rear component had become a cemented doublet; this was combined with a Tessar in front, in either an adjustable or fixed barrel.

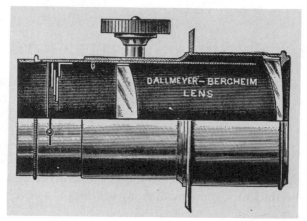

Figure 9.5. The Dallmeyer-Bergheim lens of 1893.

---

[13] U.S. Pat. 465,409 (1891).

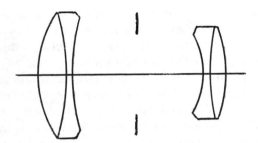

Figure 9.6. The Busch Bis-Telar $f/8$ telephoto.

In 1893 with the collaboration of the photographer J. S. Bergheim, the Dallmeyer Company offered for sale a soft-focus portrait lens of variable power consisting of two single uncorrected elements (Fig. 9.5). The low aperture of $f/9$ to $f/15$ must have been a severe limitation.

Finally, by the early years of the present century, manufacturers decided that the variable separation presented too many problems, and they began to fabricate telephoto lenses at a fixed separation, the short overall length being still a worthwhile feature in the relatively long focal lengths then in common use. The first of these was the Busch Bis-Telar lens designed by K. Martin in 1905 (Fig. 9.6).[14] The aperture was $f/7$ and the angular semi-field was 15°. This was followed by numerous others made by almost every lens manufacturer. A partial list follows:

Bausch & Lomb: Telestigmat
Dallmeyer: Adon, Dallon, Grandac
Goerz: Telegor, Telestar
Rodenstock: Rotelar
Ross: Teleros, Telecentric
Taylor-Hobson: Cotal, Eltic, Telekinic, Telic
Voigtländer: Dynaron, Telomar
Wray: Plustrar
Zeiss: Magnar, Telikon

Many companies have manufactured long-focus lenses having the prefix tele before their names, such as Tele-Tessar, Tele-Megor, Tele-Xenar, etc. In some instances, a lens has been called a telephoto when it was

---

[14] Brit. Pat. 23,888/04.

actually a normal lens with an exceptionally long focus and covering a narrow field. This practice has been common in the lenses for small movie cameras, where the shortening of the total length in a genuine telephoto by only a few millimeters would be insignificant.

The pincushion distortion of a telephoto lens has always been a problem, and at one time it was considered an inherent property of lenses of this type. In 1923 H. W. Lee of the Taylor-Hobson Company showed that the distortion could be greatly reduced or eliminated entirely by the introduction of a strong collective interface in the rear component or its equivalent, an airspace with a biconcave shape. Today's telephotos are invariably corrected for distortion along with the other aberrations.

With the progressive miniaturization of everything which began before World War I and continues today, focal lengths have become shorter and the need for telephoto lenses has gradually disappeared. Telephotos reappeared briefly during World War II for aerial cameras using formats of $9'' \times 9''$ or $9'' \times 18''$, where any shortening in the total length would be an advantage, but after the war these quickly disappeared. Today, with the almost universal use of cameras with interchangeable lenses, any lens having a focal length greater than about 135 mm is likely to be a telephoto. For very long foci, such as 500 mm or 1000 mm, mirror systems are becoming common (see Chapter 12, IIB). These may be regarded as extreme telephotos with a telephoto ratio of only 20 or 25 percent.

An interesting recent development is the wide-angle telephoto lens made by Kodak for the Disc camera (Fig. 9.7). This lens consists of a front assembly of positive elements and a large negative meniscus element near

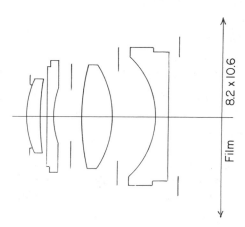

Figure 9.7. The lens on the Kodak Disc camera.

the film, the concave surface facing the front. This negative element acts as a field flattener, and it also causes the rear nodal point to move forward to yield a compact system. This lens is capable of covering an angular field approaching $\pm30°$ at $f/2.8$. In the Kodak design, a small residual of spherical aberration was removed by the insertion of a thin negative aspheric element in the front positive component. Several patents covering lenses of this general type have appeared recently, and in some of these designs the front surface of the large rear negative element is made aspheric. In some designs, the rear negative component is split into two, thus avoiding the need for an aspheric surface.

IV. MODERN TELEPHOTO ZOOM LENSES

Quite recently, the old-time varifocal telephoto system has been revived, mainly because of its compactness. A positive component in front and a negative component behind are moved in the same direction at different rates by a pin-and-slot mechanism in order to maintain a fixed image position, at least for a distant object. As in the case of an ordinary two-component zoom , it is virtually necessary to correct each component separately for all aberrations. The iris diaphragm is located within the front component and, in some cases, focusing on a close object is accomplished by a separate movement of the rear component (see also Chapter 11, Section IIIA).

CHAPTER 10

Reversed Telephoto Lenses

I. ADVANTAGES AND APPLICATIONS

Since the beginning of the present century, lantern slide projectionists have occasionally used an 'amplifier lens' to increase the size of the image on the screen. This consisted of a large negative element situated ahead of the anterior focus of the projection lens (Fig. 10.1). Its effect was to shorten the overall focal length of the lens and thus enlarge the projected image. This was the exact opposite of the well-known telenegative lens used behind an ordinary objective, which served to increase the focal length of the main lens.

The combination of a large negative lens in front of an ordinary lens consitutes a Reversed Telephoto objective. It has several advantages over a normal lens of the same focal length and aperture:

(a) Obtaining a short overall focal length with elements of a larger and more manageable size,

(b) Increasing the back focal distance beyond its usual magnitude, and

(c) Designing a system that is favorable for both a high relative aperture and a wide-angular field.

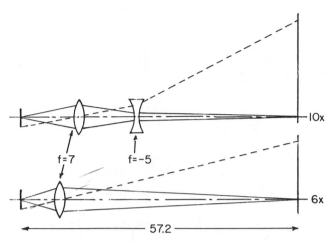

Figure 10.1. The principle of the Amplifier lens.

In view of the third advantage, we may well ask why the reversed telephoto type is not used for all applications. The reasons are that such a lens is physically large (advantage (a)), and it must be of complex construction to correct all the aberrations, making it an expensive item. It is rare to find reversed telephoto lenses over about 2 inches in focal length, and then only when this type of construction is needed for some particular reason.

Historically, lenses of the reversed telephoto type were designed about 1929 by Ball, Bowen, and others for close-up projection of motion pictures on a wide screen.[1] These lenses were a combination of an ordinary lens with an amplifier or with a reversed Galilean attachment. In 1931 H. W. Lee of the Taylor-Hobson Company designed an $f/2$ lens of 35 mm focal length for the three-strip Technicolor camera in which the beam-splitting glass cube behind the lens occupied so much space that no normal lens of focal length less than about 50 mm could be used.[2]

The earliest commercial application of reversed telephoto lenses was in the very short focal lengths required for 8 mm motion-picture cameras. A lens having a focal length of, say, 5 mm is comparable to a high-power microscope objective, yet it has to cover a wide angular field and be sold at

---

[1] R. Kingslake, "The Reversed Telephoto Objective." *JSMPTE* **75**, 203 (1966). Sectional diagrams of over 130 reversed telephoto objectives are shown in A. Cox, *Photographic Optics*, 15th ed., page 515. Amphoto, Garden City, 1974.

[2] *Phot. Journ.* **77**, 93 (February 1937). See also I. B. Happé, *Basic Motion Picture Technology*, page 27. Hastings House, New York, 1971.

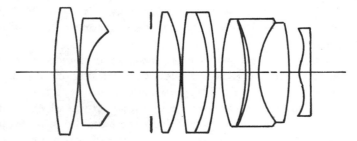

Figure 10.2. The Elgeet $f/1.2$ Golden Navitar.

a reasonable price. A large number of reversed telephoto objectives appeared after World War II for this application, many of which were quite complex; at least one involved an aspheric surface (Fig. 10.2). Also, in many of these early cameras, several lenses were mounted on a rotating turret, and in that case all the lenses had to have the same minimum back focal distance independent of their focal length.

With the advent of the 35 mm SLR camera in the early 1950s, many manufacturers began to make reversed telephoto objectives for these cameras in the shorter focal lengths. Here a back focus of at least 35 mm is required to clear the rocking mirror in the camera, and any lens with a focal length less than about 45 mm must be a reversed telephoto. Foremost in this field was the firm of Angénieux in France, which developed a line of Retrofocus lenses for this application (Fig. 10.3). The name of these lenses served only as a trademark, but this name is fast becoming a generic term for all lenses of reversed telephoto type. Such lenses for an SLR camera have a well-established series of focal lengths, namely, 35 mm, 28 mm, 24 mm, 20 mm, down to as little as 15 mm, the latter being very complex, often containing as many as fifteen elements. A typical example is the

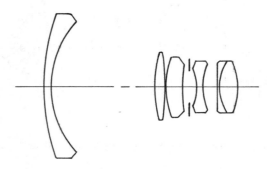

Figure 10.3. An Angénieux Retrofocus lens, 9.5 mm at $f/2.2$ (1950).

wide-angle 18½ mm lens sold by Spiratone shown in Figure 10.4. In such lenses, it is particularly difficult to correct the distortion because of their extreme lack of symmetry.

It should be remarked that the large negative element in front of a wide-angle lens of this type has the effect of increasing the illumination at the outer parts of the image. This is in direct contrast to the early wide-angle objectives in which the oblique illumination was often so low that some means had to be found to hold back the axial illumination without reducing the oblique illumination. An example of this was the spinning cogwheel used with the Goerz Hypergon lens (see Chapter 4, IIE).

A partial listing of reversed telephoto lenses is given here, but, as in previous lists, it does not follow that all lenses having this particular marker's trade name are of this type:

Angénieux: Retrofocus
Bausch & Lomb: Animar
Dallmeyer: Inverted Telephoto
Elgeet: Cine Navitar
Enna: Lithagon
Isco: Westrogon
Kern: Switar
Kinoptik: Apochromat, Tegea
Kodak: Wide-angle Ektanon

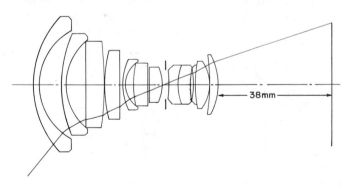

Figure 10.4. The Spiratone 18.5 mm *f*/3.5 lens. This objective covers a field of ±50°.

Leitz: all R-type objectives
Rodenstock: Eurynar, Heligaron
Schneider: Cinegon, Curtagon
Taylor-Hobson: Ivotal, Kinetal, Speed Panchro, Taytal
Voigtlander: Skopagon, Ultragon
Wollensak: Cine Raptar Wide-angle
Zeiss: Distagon, Flektogon
Zoomar: Balowstar

II. FISH-EYE OR SKY LENSES

The name fish-eye lens was coined by R. W. Wood to refer to any lens capable of imaging the entire hemisphere in object space onto a finite circle in the focal plane.[3] He chose this name because a fish looking upward at the surface of the water will see the whole sky imaged as a finite circle in this way.

Ideally, the height of the image should be proportional to the angular distance of the corresponding object point from the axis. Hence, ideally,

$$h' = f\phi$$

where ϕ is the object angle in radians. This is known as the equidistant case for a fish-eye lens or an f-θ lens when used in a scanning system with the aperture stop in front of the lens.

Fish-eye lenses generally have a very short axial focal length. For instance, if a lens of 8 mm focal length is required to cover a full hemisphere in the object space, the image height will be given by $h' = 8 \times 1.57 = 12.5$ mm, because the radian measure of 90° is 1.57. This image will just fit into the height of the standard 35 mm double frame. Most fish-eye lenses meet this condition.

Of course, there is enormous barrel distortion in a lens of this kind. If the lens were distortionless, the image height would be given by

$$h' = f \tan \phi$$

---

[3] R. W. Wood, *'Physical Optics,'* page 67. Macmillan, New York, 1911; also, Dover, New York, 1967.

The magnitude of the distortion is indicated in this table:

Semifield Angle ϕ	ϕ in Radians	Tan ϕ	Percent Distortion
0°	0	0	0
15°	0.2618	0.2679	−2.30
30°	0.5236	0.5774	−9.31
45°	0.7854	1.0000	−21.46
60°	1.0472	1.7321	−39.54
75°	1.3090	3.7321	−64.92
90°	1.5708	∞	−100.00

An 8 mm fish-eye lens has been described by designer K. Miyamoto of the Nikon Company.[4] This lens is shown diagrammatically in Figure 10.5. The Nikon Company currently makes three forms of equidistant fish-eye lens:

Focal Length	Number of Elements	F-number	Field Angle ϕ	Image Diameter	Diameter of Front Element
6 mm	12	2.8	±110°	23 mm	236 mm
8 mm	10	2.8	±90°	23 mm	123 mm
16 mm	8	3.5	±85°	46 mm	68 mm

They have also manufactured a simpler and smaller 6 mm lens having an aperture of $f/5.6$ with only nine elements.

A much less usual situation is a lens having even more barrel distortion, satisfying the orthoscopic law, where

$$h' = f \sin \phi.$$

Nikon has made such a lens with a focal length of 10 mm, which forms an image of the complete hemisphere that is only 20 mm in diameter.

The earliest example of a fish-eye lens was the Hill Sky lens of 1924, manufactured by Beck of London.[5] It was a simple device, shown in Figure

---

[4] K. Miyamoto, "Fish-eye Lens." *J. Opt. Soc. Am.* **54**, 1060 (1964).

[5] R. Hill, "A Lens for Whole Sky Photography." *Proc. Opt. Conv. 1926*, page 878; also, Brit. Pat. 225,398 (1923).

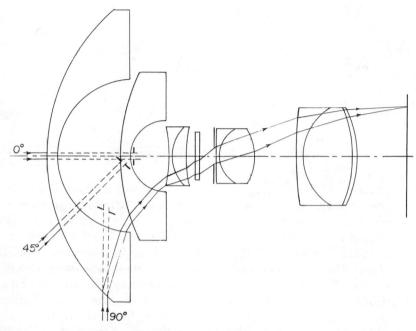

Figure 10.5. The Miyamoto design of 1964, showing the tilted entrance pupil.

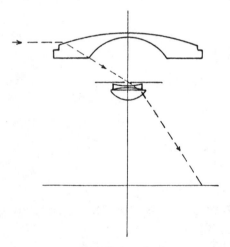

Figure 10.6. The Hill Sky lens.

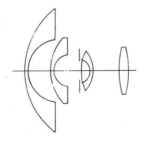

Figure 10.7. The AEG fish-eye lens.

10.6, consisting of a large negative meniscus element in front of the stop with a pair of single elements behind it. The low relative aperture was no problem to meteorologists who used the lens to photograph the whole sky on a single plate. Another more elaborate fish-eye lens was patented by the German AEG Company in 1932 (Fig. 10.7).[6] Since that time, most optical manufacturers have produced fish-eye lenses, while some have made afocal fish-eye attachments to go in front of an ordinary camera objective. The latter arrangement has the advantage that the size of the circular image can be varied by using different focal lengths for the main lens.

In any lens covering a semifield of 90° in the object space, it is obvious that the entrance pupil (the image of the diaphragm) must be tilted for the 90° beam as otherwise no light would enter the lens at that obliquity. In Figure 10.5, representing the Miyamoto lens, the position of the entrance pupil is indicated for light entering at 45° and 90° from the axis.

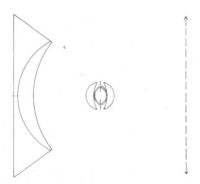

Figure 10.8. The Zeiss Pleon, for aerial photography.

---

[6] Ger. Pat. 620,538 (1932).

The presence of barrel distortion in a lens has the effect of crowding the image in the outer parts of the field and thus increasing the illumination there. There are actually four factors controlling the distribution of illumination across the field of a lens. They are:

(a) Vignetting: trimming of oblique beams by the lens apertures.
(b) The $\cos^4$ law: the edges of the field are further from the lens than the middle.
(c) Stop distortion: in some lenses, the entrance pupil diminishes in size, while in other types it increases, for oblique beams.
(d) Image distortion: pincushion distortion reduces the illumination in the outer parts of the image, while barrel distortion increases it.

In the absence of all other effects, it can be shown that the illumination across the image will be uniform if the image height follows the sine law:

$$h' = f \sin \phi.$$

In all other cases, image distortion will affect the distribution of illumination, but barrel distortion is always a help and may completely overcome the adverse effects of the other three factors listed earlier.

It should be noted that the depth of field of a fish-eye lens is very great because of its short focal length. For example, in an 8 mm $f/2.8$ lens, the diameter of the entrance pupil is just under 3 mm (⅛ inch), and the camera can be placed very close to the object without loss of sharpness. The perspective of a fish-eye lens is also unusual, an ordinary living room appearing extremely deep in such a photograph.

During World War II, the Zeiss Company developed a wide-angle aerial camera lens known as the Pleon (Fig. 10.8).[7] This lens covered a field of $\pm 65°$; it had considerable barrel distortion, which was removed by printing the negative in a special distorting printer. The story has been told by Gardner.[8] Actually, the image height in the Pleon followed the fish-eye law rather than the sine law, so the illumination was not quite uniform, but it was much better than if the lens had been distortionless.

---

[7] Ger. Pat. 722,519; U.S. Pat. 2,247,068 (1938).
[8] I. C. Gardner and F. E. Washer, "Lenses of Extreme Wide Angle for Airplane Mapping." *J. Res. NBS* **40**, 95 (1948), Res. Paper 1858.

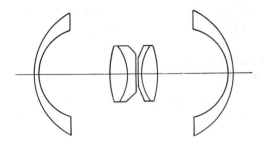

Figure 10.9. The Roosinov lens.

III. THE BIOGON TYPE

One of the most important new types of lens to be developed since World War II is a kind of double-ended reversed-telephoto objective, consisting of a compact central positive structure with one or more large negative menisci at each end, making a roughly symmetrical arrangement. The back focal distance is so short that the lens cannot be used on an SLR camera, but it has often been used on a rangefinder camera and on many aerial cameras where freedom from distortion is mandatory.

The original suggestion for this type of construction (Fig. 10.9) came from a 1946 patent by the Russian optician M. M. Roosinov.[9] Roosinov found that the use of this construction had the effect of greatly increasing

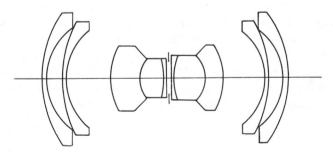

Figure 10.10. The Wild Aviogon.

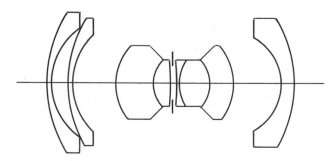

Figure 10.11. The Zeiss Biogon (second form).

the oblique illumination, even out to 66° from the axis, by causing an enlargement of the pupils at high obliquity angles. In 1952 L. Bertele decided to make use of this principle for a wide-angle aerial camera lens called the Aviogon,[10] which was manufactured by the Wild Company of Heerbrugg in Switzerland (Fig. 10.10). It was first made in a focal length of 115 mm to cover a picture 18 cm square. The distortion was less than 10 microns at any point in the field, and the definition at $f/4.5$ was excellent. The new lens quickly replaced the Topogon and Metrogon as the standard lens for aerial photography and photogrammetry.

In 1951, the Zeiss Company needed an extreme wide-angle lens for the Contax and Hasselbad cameras, so they commissioned Bertele, who had formerly been a Zeiss designer, to develop a suitable lens for this purpose. The result was the famous Biogon shown in Figure 10.11.[11] It resembled the Aviogon, but it was simplified by having only one negative element in the rear. The lens was physically large, two focal lengths in length and one focal length in diameter, but these dimensions were not objectionable when the lens was constructed in the shorter focal lengths, such as 25 mm for the Contax camera. Bertele also patented several other lenses of this general type, including some having three negative menisci at each end, which covered about 120° of total field. He could not obtain a master patent on the use of a single meniscus at each end as Roosinov had already covered this case, so other designers were free to use it. For example, Schneider designed several Super Angulon lenses of this type, some of which were used by Leitz around 1960 (Fig. 10.12).[12]

---

[10] M. Kreis, in *Festschrift, Dr. h.c. L. Bertele*, p. 11. Wild, Heerbrugg, 1975; also, U.S. Pat. 2,734,424; Brit. Pat. 680,185 (1950).

[11] U.S. Pat. 2,721,499; Brit. Pat. 719,162 (1951).

[12] U.S. Pat. 2,897,725; Brit. Pat. 830,713 (1957).

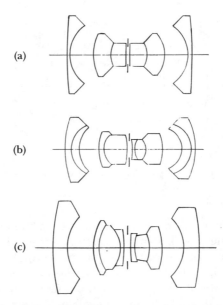

Figure 10.12. Three Schneider Super Angulon designs: (a) $f/8$, 1955; (b) $f/4$, 1958; and (c) $f/3.4$, 1963.

An interesting modification of this type is the Hologon extreme wide-angle lens announced by Zeiss in 1966.[13] It consists of only three elements (Fig. 10.13), a central positive ball surrounded by two large, thick negative menisci. Alternatively, this lens can be regarded as a reversed Triplet in which the positive element is in the middle and the two negative elements outside, as was indicated in Chapter 7, Section I. At an aperture of $f/8$, this lens covers a flat field of $\pm 55°$ without distortion, its focal length for use on a 35 mm camera being only 15 mm. Its back focal distance is too short for use on an SLR camera.

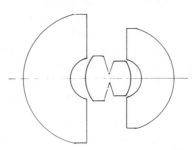

Figure 10.13. The Zeiss Hologon, aperture $f/8$, covering $\pm 55°$.

---

[13] Ger. Pat. 1,241,637 (1966).

CHAPTER 11

Varifocal and Zoom Lenses

I. INTRODUCTION

A varifocal lens is one in which the focal length can be continuously varied by moving one or more of the lens elements along the axis. In such a system, the image position generally moves with a change in focal length, but if some arrangement, optical or mechanical, is provided by which the image can be held fixed, then the system becomes a true zoom lens. In Germany, this was formerly called a rubber lens. The early history of the zoom lens has been told in many places.[1]

Obviously, a simple varifocal lens can be used on a projector or a still camera because the operator has time to bring the image into focus after selecting the desired focal length, but in a motion picture or television camera, a true zoom is mandatory, as the operator often wishes to vary the image size while the camera is running. Furthermore, in the latter case it is

---

[1] R. Kingslake, "The Development of the Zoom Lens." *JSMPTE* **69**, 534 (1960). See also H. Naumann, *Kinotechnik* **15**, 307 (1933), and *Techniksgeschichte* **33**, 279 (1966). A hundred sectional diagrams of zoom lenses are given in A. Cox, *Photographic Optics,* 15th ed., page 541. Amphoto, Garden City, 1974.

essential that the zoom condition be accurately maintained even when the object is quite close to the camera. This can be accomplished by mounting a focusing collimator[2] in front of the lens or by moving the normally fixed front component, leaving the zooming portion of the lens unchanged. Some recent zoom lenses intended for use on an SLR camera are true zooms only with a very distant object, but the entire system is moved for focusing and then it becomes merely varifocal. However, this presents no real problem because the operator invariably checks the focus before making an exposure. In a few modern zoom lenses intended to be used on an autofocus camera, focusing is accomplished by a movement of a small lens element in the rear.

In most of the early zoom lenses the aperture stop was located behind the last moving member so that the effective F-number would remain constant during a zoom. However, as most cameras today are equipped with some means for automatic exposure control, it is really immaterial if the F-number of the lens does vary somewhat, so now the diaphragm can be located at any convenient place in the system.

The component of a zoom that is used to vary the focal length is often called the variator, while some other component that is cam-operated to maintain a fixed image position is known as the compensator. Some zoom lenses made in the early 1950s embodied two separate components that were coupled together so as to move as a unit with a fixed component situated between them. This arrangement is known as optical compensation and no cam is needed (see Section IV).

In designing a zoom lens, the principal problem is to maintain good aberration correction at all times. As the airspaces in most lenses are extremely sensitive, a change by only a few tenths of a millimeter often leading to unacceptable amounts of aberration, one may well ask how it is possible to create a zoom lens at all. The secret is to arrange matters so that no aberration in any lens component ever becomes large. In two-component lenses, it is generally necessary to correct each of the components fully for all aberrations. In lenses having three or four moving components, it is often possible by careful design to ensure good overall correction by playing off the aberrations of one element against those of another. This is not easy, and most zoom lenses will be found to contain anything from as few as seven to as many as eighteen elements. However, thanks to

---

[2] A focusing collimator consists of a pair of positive and negative lenses having the same focal length that neutralize each other when in contact but develop positive power when separated.

the computer optimization programs now available, most zoom lenses yield an image quality close to that of a normal lens having the same focal length and aperture. Also, these zooms are becoming smaller and more compact all the time.

Ideally, the movement of the control lever, crank, or knurled ring that causes a variation in the focal length should be linearly related to the logarithm of the focal length, so that the user will see in the viewfinder an equal relative increase in image size for equal movements of the control. However, this condition is not always observed, and in many zoom lenses the movement of the control varies approximately linearly with the change in focal length.

It should be remarked that some form of reflex viewfinder is necessary when a zoom lens is used. In some of the older movie cameras lacking such a finder, a second zoom system was coupled to the camera zoom and provided with an eyepiece so that the operator could watch the action and the variation in image size while it was occurring.

II. HISTORICAL

Lenses with a variable focal length have been made for almost a century. The first commercial varifocal objectives were the variable telephotos made by Dallmeyer and others in the 1890s, as discussed in Chapter 9, Section II. These were not true zoom lenses, as the bellows extension of the camera had to be readjusted for each separation between the lens components.

In 1901 C. C. Allen patented a lens having a movable negative component located between the two positive components of a Petzval-type objective.[3] This again was a varifocal system; however, it turned out to be the forerunner of many current zoom designs.

In 1920 a Hollywood cameraman named Joseph B. Walker suggested a different arrangement, namely, a large negative element in front and a regular positive camera lens behind, thus making a variable-power reversed telephoto objective. Walker had been using a large negative lens as the viewfinder on a motion-picture camera and, noticing how sharp the image was, it occurred to him to focus the camera on this image. This served to widen the field, and he soon found that by varying the distance of the negative lens from the camera he could vary the size of the image on

---

[3] U.S. Pat. 696,788 (1901).

the film.[4] Walker referred to this as a Traveling lens, and he used it to make some successful motion pictures. He patented this arrangement in 1929.[5]

In 1931, Helmut Naumann of the Busch Company developed a series of variable-focus projection lenses for professional motion pictures called the Neo-Kino. In these lenses, two or three positive components were mounted in a tube and moved at different rates by a pin-and-slot mechanism to vary the focal length. This arrangement enjoyed a brief popularity as the Magnascope, used for dramatic effects in motion-picture theaters. Naumann followed this by a zoom lens for 16 mm movie cameras called the Vario Glaukar (see Section IIIB).

In the following year, Arthur Warmisham of the Taylor-Hobson Company developed an elaborate zoom lens for 35 mm motion-picture cameras called the Varo.[6] It was contained in a heavy aluminum box with three moving components, an achromatic negative in front, a high-aperture double-Gauss in the middle, and another negative achromat in the rear, all of which were made to move at different rates by turning a crank (Fig. 11.1). The system thus combined a reversed telephoto zoom system in front and an ordinary telephoto zoom in the rear. The large central positive member and the rear negative component were moved linearly at different rates, while the front negative component was moved in and out by a cam to hold a constant focal plane. Focusing on near objects was accomplished by adding a weak collimating lens in front, which, of course, did not affect the zoom law. The focal length of this system ranged from 40 mm to 120 mm at $f/3.5$. Because the diaphragm was located inside the middle component, it had to be driven by a cam during a zoom in order to maintain a fixed relative aperture, a different cam being required for each F-number.

Because of the great Depression that raged during the 1930s, no further development of zoom lenses was undertaken until after World War II. In 1945, Frank Back produced the first Zoomar lens (see Section IV), an optically compensated structure for 16 mm and 35 mm movie cameras. This initiated a rapid development of zoom lenses for both motion pictures and television.

Actually, professional motion-picture cameramen were somewhat re-

---

[4] J. B. and B. Walker, *The Light on Her Face*, page 266. ASC Press, Hollywood, 1984. See also *JSMPTE* **94**, 1210 (1985).

[5] U.S. Pat. 1,898,471.

[6] A. Warmisham and R. F. Mitchell, "The Bell & Howell-Cooke Varo Lens." *JSMPE* **19**, 329 (1932). Also Brit. Pat. 398,307; U.S. Pat. 1,947,669 (1931).

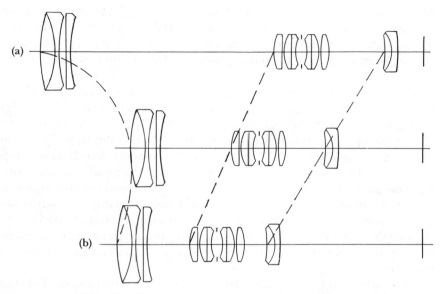

Figure 11.1. The Cooke Varo lens: (a) $f = 40$ mm and (b) $f = 120$ mm.

luctant at first to use zoom lenses. This may have been due to conservatism or to the unfamiliar perspective effects that were not the same as when the camera is moved toward or away from the actors. It took the universal adoption of zoom lenses by the infant television industry to overcome this prejudice.

One of the earliest TV zoom lenses was the Electrozoom, developed by Walker following the success of his Traveling lens of the 1920s. This name was chosen because a small electric motor was used to drive the two moving components, the front negative element being moved rapidly while the rear positive component was moved slowly in the opposite direction. An elaborate cam mechanism was required to maintain the zoom law when used with a near object. This system covered a focal length range from 52 mm to 152 mm at $f/3$.

A different arrangement was suggested in 1951 by H. H. Hopkins for the Watson TV camera zoom lens.[7] In this system, two fixed outer positive lenses were used with a moving negative component between them (Fig.

[7] H. H. Hopkins, "A Class of Symmetrical Objectives of Variable Power," in Optical Instruments: Proceedings of the London Conference 1950, (W. D. Wright, ed.), page 17. Chapman and Hall, London, 1951. Also Brit. Pat. 685,945; U.S. Pat. 2,663,223 (1950).

11.2). However, to keep the image always in focus, Hopkins divided the negative component into two parts at a variable separation. Thus, the entire negative component became the variator while the separation between the two parts constituted the compensator. In this way, a 5 : 1 range in focal lengths was achieved with good aberration correction.

The development of zoom lenses for still cameras came much later. For many years it was generally assumed that the availability of several interchangeable lenses was all that photographers needed, as it enabled them to choose the optimum focal length for any particular application. The same idea was also applied to small amateur movie cameras, which were often fitted with a turret carrying three lenses that could be rapidly brought into position as required. (Turrets have also been proposed for still cameras, but the sheer size and weight of the lenses make this impractical.) However, the operator of a motion-picture camera would often desire to vary the image size while the camera was running; a zoom lens was then essential. Thus, for many years zoom lenses were considered to be purely a motion-picture problem.

Although photographers were free to carry a bag full of lenses wherever they went, it was a cumbersome and expensive arrangement, and by the early 1960s they were beginning to ask why there were no zoom lenses available to them. The problems involved in designing a zoom lens for still

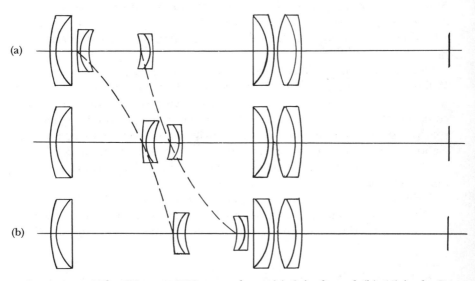

Figure 11.2. The Watson TV zoom lens: (a) 3 inch and (b) 15 inch. By changing the rear element, the focal length and F-number can be doubled.

and movie cameras are quite different. The main difference is one of scale. The diagonal of a Super-8 frame is 7.1 mm, while that of a 35 mm still camera is 42 mm, about six times as large, yet the two systems must be comparable in physical size for reasons of portability and convenience in use. If a movie camera zoom were enlarged six times, it would be huge and expensive, and the angular field would be much too narrow. Conversely, if a still-camera zoom were reduced in size by a factor of six, it would become a tiny affair with much too small a focal-length range and too low an aperture. Furthermore, as has already been emphasized, there is no need to maintain a fixed image plane in a zoom lens for a still camera. However, the many problems have been overcome, and a large number of excellent zoom lenses covering a wide range of focal lengths are now available for 35 mm SLR cameras, the image quality being as good as in a normal lens of the same focal length and aperture.

A few varifocal lenses have also been developed for motion-picture and slide projectors. These are, of course, a great convenience, as the size of the projected image can be adjusted to fill the screen from any convenient projection distance.

III. ZOOM LENS TYPES

The hundreds of zoom lenses that have been developed over the past 35 years fall mainly into four recognizable types, which are named after the number of lens groups that have an independent existence in the system and are either moved or held stationary during a zoom. The two-, three-, and four-component zoom types are mechanically compensated by a cam or pin-and-slot mechanism, and all three types are being extensively manufactured today. In addition, there have been a number of optically compensated systems that were very active in the 1950s and still appear occasionally in patent disclosures. A few excellent modern zooms for still cameras do not fall into any recognized type, but represent attempts on the part of the designer to improve the aberration correction, to reduce the size of the lens, or to provide a macro feature. Sometimes a modern zoom lens has been especially adapted for use with an auto-focus camera of some kind.

A. The Two-Component Zooms

The use of two lens components at a variable separation is an obvious way to produce a varifocal system. Indeed, the first telephoto lenses of

1891 were of this type (see Chapter 9). Quite recently some properly designed telephoto zoom lenses have been developed for use on a 35 mm camera in the longer focal lengths where the compactness is a real virtue.

A much more useful procedure has been to follow Walker's suggestion and construct a reversed-telephoto system with a negative component in front and a positive component in the rear, the positive component being normally somewhat stronger than the accompanying negative component.

Figure 11.3 shows the effects of moving the two components of a telephoto and a reversed telephoto system. The focal lengths of the components have been arbitrarily set at $+3$ and -4 units in both cases. It will be seen that the movement of the rear component varies linearly with the overall focal length, while the front component moves to a minimum position when the overall focal length is equal to the absolute value of the focal length of the front component. Thus, in the telephoto case this minimum occurs when the negative rear component is actually in the focal plane, representing a practical limit to the range of movement. It is clear

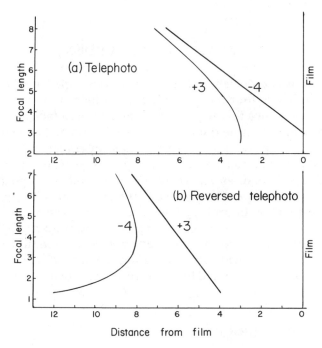

Figure 11.3. Movements of components in a telephoto and a reversed telephoto zoom.

also that the telephoto zoom systems are shorter than their focal length, and if they are to be used on an SLR camera, the back focus clearance must be greater than 35 mm to clear the rocking mirror on the camera.

The reversed telephoto arrangement shown in Figure 11.3(b) is much more interesting and more generally useful. The original lens by Walker was, obviously, located in the lower part of this graph where the lens separation is large and the focal length is short, shorter in fact than that of either component. In this lower region the movements of both components are virtually linear, as Walker had found. On the other hand, most of the later reversed-telephoto zooms have tended to be located around the minimum point to yield a compact system. The variator is then the rear component and the compensator is the front component.

Two factors must be considered in the design of a zoom lens of the two-component type. The first is that the diaphragm is generally located in one of the components and moves with it; consequently, the relative aperture changes during a zoom. However, the automatic exposure control in many modern cameras makes this of little account. The second factor is that since both components are moveable and fairly strong, it is virtually necessary to correct each completely for all aberrations; many two-component zooms contain from four to six elements in each half. With the help of a computer optimization program, this does not lead to any great difficulty.

The reversed-telephoto zoom system enjoys the benefits of all reversed telephotos in that it has a long back-focus clearance and good aberration correction for both a high relative aperture and a wide angular field. Consequently, all zoom lenses designed to cover a wide angular field are of the reversed telephoto type. Some typical examples are shown in Figure 11.4(a) and (b), representing two Pentax zoom lenses designed by Takayuki Itoh of the Asahi Company. In Figure 11.4(c) the thin-lens movements of these designs are indicated.

A few wide-angle zoom lenses have reached a short-focus limit of only 21 mm, but this is rather extreme; 24 mm represents a more practical end point. Some lenses of this type began to appear in the 1970s, and they are now quite common. The relative aperture is generally $f/3.5$ at the wide-angle end, dropping to $f/4.5$ at the long-focus end of the range.

B. Zoom Lenses Depending on the Donders Principle

Somewhere around 1880 the Dutch ophthalmologist F. C. Donders suggested a form of three-element variable power telescope of the type shown diagrammatically in Figure 11.5. In these systems, the two outer

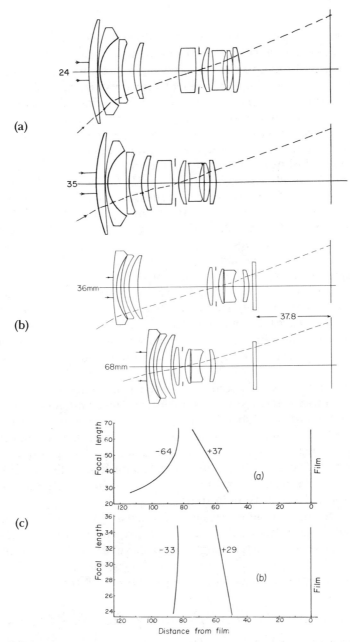

Figure 11.4. (a) Zoom lens by Itoh, U.S. Pat. 4,196,968 (1977), (b) zoom lens by Itoh, U.S. Pat. 4,726,665 (1985), and (c) movements of components in the two previous Itoh lenses.

162

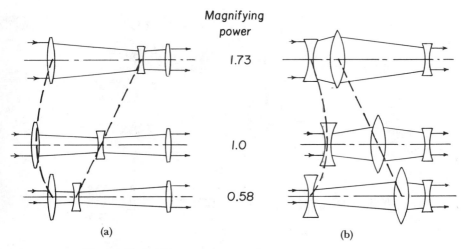

Figure 11.5. The two types of Donders telescopes.

elements are intended to be fixed, and a sliding inner element of opposite sign serves to vary the magnifying power of the telescope. By a suitable choice of the powers and separations of the lenses the system can be made afocal. If the outer components are of equal power and the inner component is midway between them, the magnifying power will be unity, and the inner unit will be working at a magnification of − 1. Two possibilities exist. In the case shown in Figure 11.5(a), when the inner component is moved forward the power will drop, while in Figure 11.5(b) the reverse applies. Provided the movement of the inner component is small the system will remain virtually afocal, but for a greater movement it is necessary to reestablish the afocal condition by a small in-and-out movement of one of the outer components.

In 1931 it occurred to Helmut Naumann of the Busch Company that such a variable-power telescope could be mounted in front of an ordinary camera lens to construct a zoom system.[8] The overall focal length of the combination would be determined by the product of the focal length of the camera lens and the magnifying power of the telescope. The paths of rays through a schematic 3 : 1 system of this type are shown in Figure 11.6, and the actual construction used by Naumann in his Vario Glaukar lens is shown in Figure 11.7. It was manufactured by Busch for use on a Siemens 16 mm movie camera in a range of focal lengths from 25 mm to 80 mm at

---

[8] H. Naumann, *Das Auge meiner Kamera*, page 116. Knapp, Halle, 1949.

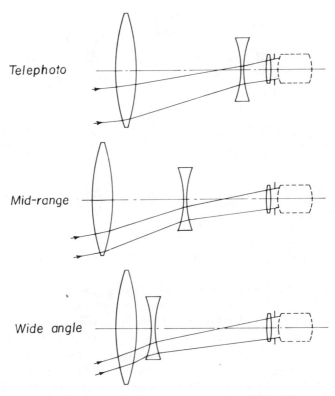

Figure 11.6. Combination of a 3 : 1 Donders telescope with a normal lens.

$f/2.8$. The front component served as both the compensator and the focusing unit for use with near objects; it moved outward at midrange and back in again at both ends of the zoom movement. This can, therefore, be regarded as a three-component zoom.

Of course, there is no magic about having the zoom attachment precisely afocal, and in many zooms this condition is not observed. However, if a beam-splitting prism is to be inserted between the zoom attachment and the main lens to reflect an image into the viewfinder, then parallel light in this space is desirable. The aperture stop and the small photocells that control the exposure are conveniently located in this space also.

It will be noticed that if the compensating lens is positive it must be moved toward the variator at each end of the zoom range, and hence the two components move toward each other at one end of the range and set a limit to the zoom motion there. In many designs, therefore, the compensa-

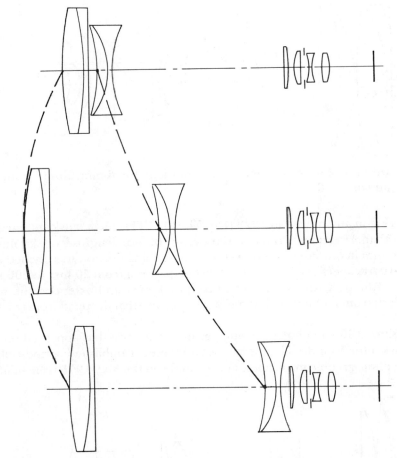

Figure 11.7. The Busch Vario-Glaukar.

tor lens is made negative so that it will move away from the variator, but it must be followed by a strong positive element to maintain the afocal condition.

Two excellent examples of this type of zoom lens are the 8 : 1 and 10 : 1 zooms developed by Angénieux for 8 mm and 16 mm movie cameras in the mid-1960s.[9] They are shown in Figures 11.8 and 11.9.

An extreme example of this type of zoom lens was the Schneider TV-

[9] D. F. Horne, *Lens Mechanism Technology*, page 31. Crane Russak, New York, 1975.

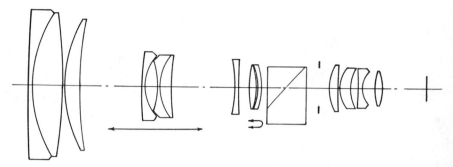

Figure 11.8. The Angénieux 8:1 zoom lens for 8 mm film, 6.5 mm to 52 mm at $f/1.8$.

Variogon announced in 1973 (Fig. 11.10).[10] This huge lens contained no less than 31 elements and covered a range of focal lengths from 20 mm to 600 mm in one continuous sweep. The zooming portion was divided into two parts operating in tandem. The first part ran from 20 mm to 200 mm at $f/1.6$, after which the second part took over and extended the focal length from 200 mm to 600 mm at an aperture that dropped from $f/1.6$ to $f/6$.

Since 1960 an enormous number of zoom lenses has appeared on the market for 35 mm still cameras. Lists of these are published periodically in the photographic magazines. For example, in the May 1975 issue of *Mod-*

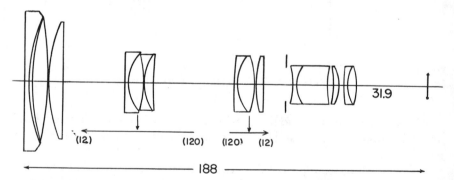

Figure 11.9. The Angénieux 10:1 zoom for 16 mm film, 12 mm to 120 mm at $f/2.2$.

---

[10] K. Macher, "Neues Kronstructionsprincip für Vario-objective mit grossem Brennwei-tenbereich." *J. Schneider Hausmitteilungen* **18**, 85 (1973).

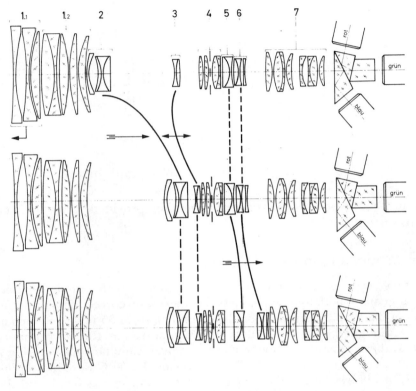

Figure 11.10. The Schneider TV-Variogon.

ern Photography there is a listing of over a hundred zoom lenses available, with focal lengths ranging from 35 mm at the wide-angle end to as much as 600 mm for a long-focus zoom, all at very high prices. Today, only fourteen years later, the range of focal lengths is greater, especially at the wide-angle end, and prices have dropped to about one-third. The size and weight of the lenses have been much reduced, and the image quality is better thanks to the availability of computer optimization programs.

The other type of Donders telescope was used by H. Gramatzki of the Astro Company in his Transfokator attachment of 1934 (Fig. 11.11).[11] However, the need for much stronger elements and the large diameter of the internal positive element restricted the range of magnifying power to

---

[11] Ger. Pats. 650,607 and 622,046 (1946).

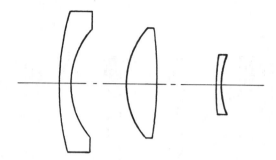

Figure 11.11. The Astro Transfokator attachment.

about 2×, and no compensating movement was provided. It has not been used since that time.

C. Modern Complex Designs

It has recently been found advantageous to modify the simple Donders arrangement, both for compactness and aberration correction. For example, in the Vivitar Series 1 zoom lens, running from 35 mm to 85 mm at $f/2.8$, designed by E. I. Betensky, which appeared on the market about 1974, three separate components were moved simultaneously during a zoom, as shown in Figure 11.12. Components B and C were moved linearly at different rates, the rear positive component was fixed, and the front component A was cam-driven to maintain a fixed focal plane for distant objects. No attempt was made to hold the image plane with a near

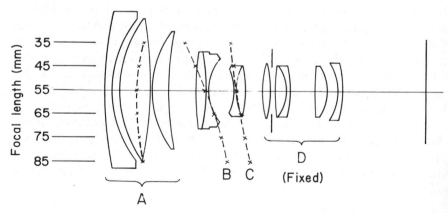

Figure 11.12. The Vivitar zoom lens, 35 mm to 85 mm at $f/2.8$.

object, and the whole system had to be focused manually by the operator immediately before making an exposure. The image quality of this lens was fully comparable with that of a normal fixed lens of the same aperture and focal length.

Ten years later, in 1984, Vivitar announced a zoom lens running from 28 mm to 135 mm at $f/3.5$ to $f/4.5$. The design of this lens is shown diagrammatically in Figure 11.13. It will be seen that this design resembles a combination of a Donders system and an optically compensated arrangement, the front and rear components being coupled together and moved forward to increase the focal length, while the second (negative) component is simultaneously moved backward. A small positive third component serves as a compensator to hold the focal plane when photographing a distant object. The relative aperture drops from $f/3.5$ at the wide-angle end to $f/4.5$ at the long-focus end of the zoom range. Some recently designed zoom lenses actually cover a 7 : 1 focal length range, from 28 mm to 200 mm.

For many years the Taylor-Hobson Company in England has been making a series of high-aperture Varotal zoom lenses for professional motion-picture and television cameras. These were designed under the direction of G. H. Cook and date from about 1950.

D. Macro Zooms

An interesting recent development is the design and fabrication of macro zoom lenses. In these systems, by turning a ring on the lens barrel the internal mechanism can be changed, some of the previously moving components becoming fixed while others may be moveable. Then, by operating the normal zoom control, the camera can be focused on very

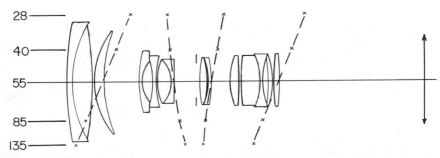

Figure 11.13. The Vivitar zoom lens, 28 mm to 135 mm at $f/3.5$ to $f/4.5$.

close objects without loss of definition. In this way the image magnification can be raised almost to unity, the object being located very close to the front lens surface.

IV. OPTICALLY COMPENSATED ZOOM LENSES

The term optically compensated implies that the image is made to remain in a fixed focal plane by optical rather than by mechanical means, without the need for any cam. At one time it was feared that a cam might become worn, leading to an unwanted movement of the image, but this fear was groundless, and optically compensated zoom lenses have virtually disappeared.

The first optically compensated zoom lens was the Zoomar, designed by Frank Back in 1946.[12] It was a long narrow system with a focal-length range from 17 mm to 53 mm at $f/2.9$, for use on a 16 mm movie camera. The lens contained 22 elements arranged as shown in Figure 11.14. There were five groups of elements, in succession: an objective, a moveable field-lens system, a fixed erector, a second moveable field lens, and a final relay. The two field lenses were coupled together and moved as a single unit during a zoom. The final relay was needed to shift the image away from the second field lens and also to invert the image on the film. It is not known how Back determined the powers and separations of the five units in this system to produce a reasonably fixed final image nor how he determined the actual structures of the various components. The original Zoomar was made in two sizes, for 16 mm and 35 mm film. In normal use a second identical zoom was mounted alongside the camera lens and coupled to it; this second zoom was equipped with an eyepiece for use as a viewfinder and yielded an erect image.

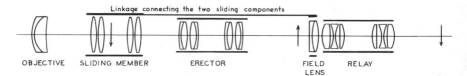

Figure 11.14. The original Zoomar lens.

---

[12] U.S. Pat. 2,454,686; Ger. Pat. 815,110 (1946).

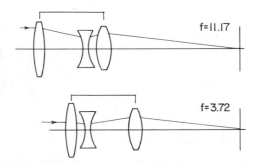

Figure 11.15. A three-element optically compensated zoom.

The next optically compensated zoom system to be developed was the Pan Cinor, designed in 1949 by R. H. R. Cuvillier of the SOM-Berthiot Company in France.[13] In this system two positive lenses were coupled together and moved as a single unit with a fixed negative lens between them (Fig. 11.15). With some care it is possible to determine the powers and separations of the three components so that the image plane remains relatively fixed during a zoom. The theory has been given in many places,[14] and it is shown that the image actually moves slightly forward and backward during a zoom, there being three positions at which the image returns to its ideal plane, with small loops or excursions of the image between them. The magnitude of the loops can be considerably reduced

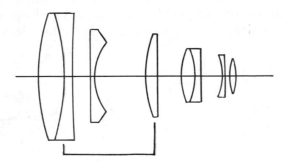

Figure 11.16. The SOM-Berthiot Pan Cinor, 20 mm to 60 mm at $f/2.8$, for 16 mm film.

---

[13] French Pat. 983,129; U.S. Pat. 2,566,485; Brit. Pat. 668,125 (1949).

[14] R. Kingslake, *Lens Design Fundamentals*, p. 66. Academic Press, New York, 1978.

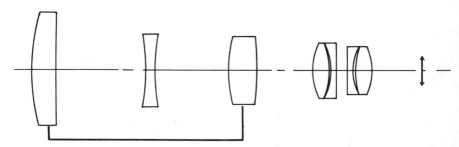

Figure 11.17. The Wollensak projection zoom lens, 15 mm to 25 mm at $f/1.5$ (1960).

by adding a strong fixed objective behind the zoom system; the zooming portion can then be made as large or as small as convenient. The actual structure of the original Pan Cinor lens is shown in Figure 11.16, and all the later Zoomar lenses were constructed on this principle. Actually, the lens elements can be made quite simple if the focal-length range is small, as indicated in Figure 11.17. Focusing an optically compensated lens can be accomplished by merely varying the separation between the coupled moving elements.

A much better way to reduce the size of the loops is to construct a five-lens arrangement consisting of two fixed lenses of one sign with a pair of coupled moving lenses of opposite sign between them. Such a five-lens zoom system is shown in Figure 11.18. There are now four zoom positions at which the image returns to its ideal position, and the loops between them are very small indeed. A comparable pair of $f/4$ three-lens and five-lens zooms are:

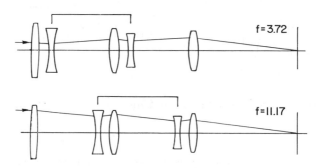

Figure 11.18. A five-element optically compensated zoom.

	Three Lenses		Five Lenses	
	f	d	f	d
	7.37	3.00	15.84	1.00
	−2.07	1.09	−4.89	4.00
	3.15		5.12	1.00
			−4.89	4.00
			4.58	
Initial back focus	8.46		6.45	
Initial focal length	11.17		3.72	
Movement of coupled lenses	2.00		3.00	
Final focal length	3.72		11.17	

These two systems are drawn to the same scale in Figures 11.15 and 11.18. It is clear that all the elements in the five-lens case are weaker than those in the three-lens arrangement, which accounts in part for the smallness of the loops there. Actual computation shows that the loops in the 3-lens case are respectively +0.066 and −0.037, whereas those in the 5-lens case are +0.0009, −0.0008, and +0.0018, which are of the order of one-fiftieth of those in the three-lens arrangement. For the same focal-length range, the five-lens case is physically much larger; however, it could obviously be made to work satisfactorily over a longer range than the 3 : 1 shown here.

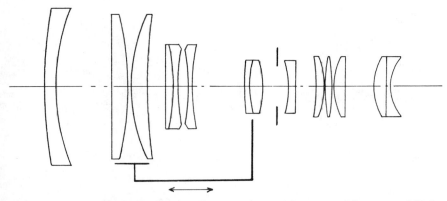

Figure 11.19. The Voigtländer-Zoomar lens, 36 mm to 82 mm at $f/2.8$.

The first practical application of the five-lens zoom arrangement was an SLR camera zoom designed by Frank Back in 1958 and manufactured by Voigtländer in Germany.[15] Because the moving lenses were positive, the zoom portion formed a real image, but the focal length was so long that it was necessary to add a positive component in the rear, as shown in Figure 11.19. By the use of fixed negative elements the angular field could be made wider, and this lens actually covered a range of focal lengths from 36 to 82 mm at $f/2.8$. The stop was located in front of the fixed rear portion, so the relative aperture did not change during a zoom. The only real problem with this lens was an unacceptable amount of distortion at the ends of the zoom range.

---

[15] U.S. Pat. 2,902,901 (1958).

CHAPTER 12

Catadioptric (Mirror) Systems

The ancient prefixes dia and cata have recently been revived to refer to optical systems in which the light is, respectively, transmitted or reflected. If both refracting and reflecting surfaces are used in the same system the name catadioptric is used; if only mirrors are involved the system is called catoptric.

I. CATOPTRIC SYSTEMS

The first mirror system to be used for photography was the daguerreotype camera patented by A. S. Wolcott of New York in 1840.[1] This camera consisted of a concave mirror, probably made by Henry Fitz, about 7 inches in diameter and some 12 or 14 inches in focal length, mounted at the back of a closed box, with a daguerreotype plate 2 × 2½ inches in size near the front end of the box facing back toward the mirror. It is hard to believe that the definition would be acceptable with such a simple uncor-

---

[1] U.S. Pat. 1,582 (1840).

rected spherical mirror, but undoubtedly many portraits were made by its use both here and in Europe.

This camera had the advantage of a high relative aperture, $f/2$, and it also formed a correct image instead of the left-handed images produced by an ordinary daguerreotype camera. However, the smallness of the picture counted against it, and the camera soon disappeared. The only known example is in the York County Museum at Saco, Maine, and even this lacks its mirror.

II. CATADIOPTRIC SYSTEMS

Since about 1930 a number of reflective systems have been developed for photography, some of which employ a single mirror together with such lenses as are needed to correct the aberrations, while others contain two mirrors, one concave and one convex, in a Cassegrain arrangement, leading to great compactness. This system is analogous to a telephoto lens having a telephoto ratio of only 15% or 20% instead of the 80% or 90% common in an ordinary telephoto lens (Fig. 12.1).

A. Single-Mirror Systems

When used with a distant object, a spherical mirror suffers from spherical aberration, and it has long been known that this can be removed by making the mirror parabolic rather than spherical, as in many astronomical telescopes. Unfortunately, the angular field of a parabolic mirror is

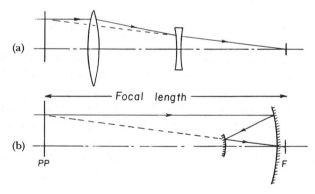

Figure 12.1. Comparison of a telephoto lens and a Cassegrain mirror system, both at $f/4$: (a) an 80% telephoto and (b) a 25% Cassegrain.

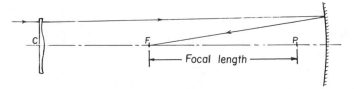

Figure 12.2. An $f/3$ Schmidt camera. (The asphericity is greatly exaggerated.)

extremely narrow, only a few minutes of arc, and such a mirror is unsuitable for ordinary photography. Consequently, many efforts have been made to use a spherical mirror, both for simplicity in manufacture and also because of its much wider field.

The first successful step in this direction was made in 1932 by the Estonian optician Bernhard Schmidt.[2] Schmidt realized that if a stop is mounted at the center of curvature of the spherical mirror the oblique aberrations coma and astigmatism will be automatically eliminated and by placing a thin aspheric plate in the stop plane he could remove the spherical aberration also (Fig. 12.2). In this way Schmidt was able to increase the angular field to several whole degrees, and since his time many large Schmidt cameras have been constructed for astronomical purposes. It should be mentioned that the focal surface of the Schmidt camera is spherical, concentric with the concave mirror. The problems involved in fabricating the aspheric plate are difficult and limit this procedure to large astronomical instruments.

About 1941, it occurred to several workers that one or more weak negative lenses could be used instead of an aspheric plate. For instance, Gabor[3] and Houghton[4] proposed inserting a negative system in front of the concave mirror, shaped so as to correct the spherical aberration and located at such a position as to remove the coma. In this way they designed practical camera systems that would cover several degrees at quite a high relative aperture, approaching $f/1$ in some cases. By careful design the chromatic aberration could also be corrected. By the addition of a positive field flattener lens close to the focal surface, the image could be flattened and the astigmatism corrected also.

---

[2] *Sky and Telescope* **15**, 4 (November 1955).
[3] Brit. Pat. 544,694 (1941).
[4] Brit. Pat. 546,307; U.S. Pat. 2,350,112 (1941).

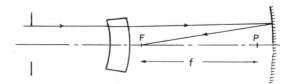

Figure 12.3. An $f/3$ Bouwers-Maksutov system.

In an effort to increase the angular field still further, it occurred simultaneously to A. Bouwers in Holland[5] and D. D. Maksutov in Russia[6] that it is possible to use a form of correcting lens in which both surfaces are spherical and concentric with the mirror (Fig. 12.3). Such a system is monocentric and has no specific optical axis; the angular field is therefore unlimited in extent, but the presence of zonal spherical aberration sets a limit to the relative aperature. It turns out that if the corrector lens is close to the stop, it will be deeply meniscus in shape, and it must be made very thin to introduce the required amount of spherical overcorrection. On the other hand, if the corrector lens is made thicker it must be placed further from the stop, and it will then contribute less zonal aberration but it will be heavier. Indeed, the weight of a Maksutov corrector lens tends to set a limit to its usefulness. There will also be some degree of chromatic aberration in these monocentric systems, and the field will be concentric about the center of curvature of the system. Ordinarily, the focal point is located inside the system, but it is possible to insert a plane mirror to reverse the direction of the light and form an external image at a much more convenient location; unfortunately, such a mirror is liable to cause an increase in the obstruction. Maksutov systems have occasionally been used for projection television, a small CRT with a curved front face being mounted inside the system, with no additional plane mirror or field-flattener lens.

B. Two-Mirror Systems

To make a practical photographic device it is necessary to add a second, usually convex, mirror so that the emerging beam will be in the same direction as the entering beam (Fig. 12.4). Such systems are generally known as Maksutov-Cassegrain systems, and many types have been designed and manufactured. They are especially convenient in the longer

---

[5] A. Bouwers, *Achievements in Optics*. Elsevier, Amsterdam, 1946.
[6] *J. Opt. Soc. Am.* **34**, 270 (1944).

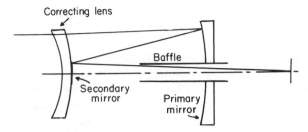

Figure 12.4. A typical Maksutov-Cassegrain system.

focal lengths, such as 400 mm to 2000 mm for an SLR camera, where they act like an extreme telephoto lens, and where the narrow angular field of this type is no disadvantage. Such systems have been offered for sale by Askania, Old Delft, Questar, Vivitar, Wollensak, Zoomar, Zeiss, and other companies. In many of these instruments the secondary mirror consists of an aluminized disk in the center of the rear convex surface of the corrector lens.

It should be noticed that a catadioptric system of this kind cannot be equipped with a normal iris diaphragm because of the large central obstruction. This obstruction inevitably causes some loss of definition, which would quickly become worse if the aperture were reduced still further by an iris diaphragm. The exposure time must, therefore, be determined entirely by the shutter speed. Focusing on close objects is accomplished by reducing the space between the two mirrors. For terrestrial use in daylight, as opposed to astronomical applications, it is necessary to provide suitable baffles to prevent ambient light from falling directly on the film. These baffles are indicated in Figure 12.3.

The Vivitar Company recently offered for sale a catadioptric lens called the Solid Cat, manufactured by Perkin and Elmer, in which the space

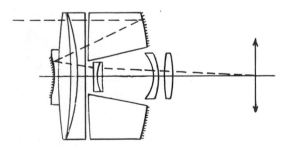

Figure 12.5. The Vivitar Solid Cat, 600 mm at $f/8$.

between the mirrors was filled with glass (Fig. 12.5). This added seriously to the weight of the system, and most manufacturers now prefer the ordinary airspaced arrangement.

Because of the central obstruction, the little catch-lights appearing on shiny surfaces in a catadioptric photograph are doughnut-shaped instead of being small circles of light. The effective F-number of such a unit is determined by the square root of the difference between the squares of the F-number of the outer rim of the aperture and the F-number of the circular obstruction. Thus, a system with an exterior aperture of $f/4$ and an obstruction of $f/8$ has an effective aperture of $f/4.62$. Hence, even a 50 percent obstruction does not greatly reduce the effective speed of the system.

CHAPTER 13

Lens Attachments

Many different forms of attachments to lenses have been developed. Some of these are intended to be added in front of a camera lens, to permit focusing on a close object or to change the focal length of a lens. Others have been added behind the lens, to affect the relative aperture or the focal length, or to flattten the field. Each of these attachments will now be considered.

I. FRONT ATTACHMENTS

A. Filters

While not in themselves lenses, filters have long been used for various purposes. Color filters have obvious uses. Neutral density filters have been used to reduce the image illumination when a very fast film is in the camera or to hold back the skylight while allowing all the light from the foreground to pass. A filter with a neutral density spot in the center has been used to reduce the axial illumination in cases of severe vignetting. Polarizing filters are used to darken the sky without affecting the brightness of clouds or to eliminate unwanted reflections from glass or other shiny nonmetallic surfaces.

Another type of nonoptical front attachment is the lens hood, used mainly to prevent light from the sun or other bright source from entering the lens. A well-placed hood can often prevent the formation of ghost images or flare spots.

B. Diffusion Disks

Another common type of filter is the diffusion disk. This usually consists of a plate of glass on which a pattern of shallow grooves has been generated. Sometimes veins or ribs of higher-index glass are imbedded in lower-index glass for the same purpose. The result is a sharp image covered with a haze of unfocused light; this was once considered desirable in portraiture, particularly in the period between the two World Wars.

C. Diopter Lenses

Most optical manufacturers offer sets of simple meniscus attachment lenses in a series of diopter powers. Positive lenses of one, two, and three diopters power are the most common, but occasionally negative attachments of these powers are offered. A positive diopter attachment enables a fixed-focus camera to be focused on a near object; it can also be used to increase the focusing range of a focusable lens. Negative diopter lenses are useful only on a bellows-type camera, where they have the effect of moving the focal plane away from the lens and thus causing the image to become larger.

Because diopter lenses are uncorrected for all aberrations, their use can lead to some loss of definition. This is generally insignificant with a one-diopter lens, but it may become noticeable if a three-diopter attachment is used. If very small objects are to be photographed, diopter lenses are helpful, but it is generally better to use a small bellows or other means for increasing the separation between the lens and the camera.

D. Afocal Attachments

It appears that the first afocal front attachment to be proposed was the Dallmeyer Adon lens.[1] This was a small Galilean telescope, and it served to increase the focal length of the main lens by a factor equal to the magnifying power of the attachment. However, it was soon found that by increas-

---

[1] Brit. Pat. 24,720/99; U.S. Pat. 756,779; Ger. Pat. 120,480.

ing the separation between the front and rear components of the Adon it would project a real image without the need for a basic lens, and it was then sold as a normal telephoto objective.

Afocal Galilean attachments became available for 8 mm movie cameras in the late 1950s. The plan was to provide a rotatable turret carrying two Galilean attachments, one the right way round to give a magnification of about 1.6×, while the other was a reversed Galilean with a magnification of about 0.6× (Fig. 13.1). A third hole in the turret enabled the main lens to be used alone without any attachment. The advantage of this scheme over the use of three normal lenses was that the attachments need not be carefully centered or accurately spaced from the main lens as the parallel light between the attachment and the lens made this precision unnecessary. A few attempts have been made to develop Galilean attachments for 35-mm still cameras, but their large size and small magnification ratio made them unpopular and expensive.

It should be noted that because the iris diaphragm is inside the main lens, the use of an afocal attachment has no effect on the F-number, but the focus scale of the camera is drastically altered. In a few instances the attachment itself could be focused on a near object and it then carried its own distance scale. The coming of zoom lenses for these cameras quickly rendered these Galilean attachments obsolete.

A few Galilean attachments have been manufactured for use on the lens in an 8 mm or 16 mm film projector. In one case the same attachment could be added either way round, to increase or reduce the size of the projected image as required.

A few companies have made afocal fish-eye attachments for use on the lens of an SLR camera (see Chapter 10, Section II).

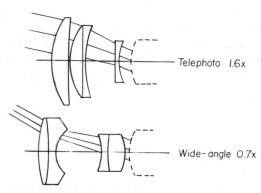

Figure 13.1. Afocal Galilean attachments for a small movie camera.

E. Anamorphic Attachments

If an afocal attachment is constructed entirely of cylindrical surfaces with their cylinder axes parallel, the focal length of the main lens will be changed only in the power meridian of the cylinder lenses, with no effect in the null meridian. The result is, of course, some degree of anamorphic distortion, with the conversion of a square object into a rectangular image or the conversion of a circle into an ellipse. This application was patented by P. Rudolph in 1898.[2]

Anamorphic compression has found only a few applications on still cameras or enlargers, mainly in advertising or the printing industry, but it has been enormously successful in the motion-picture field. In 1910 Ernesto Zollinger of Turin, Italy, took out an American patent, the first claim of which read:

> "The art of producing motion pictures, which comprises deforming the picture on the film by reducing one of its dimensions to a fraction thereof and projecting the deformed picture through a deformer to reconstruct the projection to normal proportions."[3]

Actually, Zollinger's aim was to reduce the frame height to half-size to save film. In 1928 H. Chrétien (1879–1956) patented an identical arrangement, but he placed the power meridian horizontally, so that a long narrow scene could be compressed to the normal picture format in the camera and then expanded to its original shape on projection.[4]

Chrétien's process was demonstrated before the Optical Society of America at its Washington meeting in October 1928, but it roused no particular interest, probably because it was shown on too small a scale.[5] However, the idea caught on, and in 1930 the English Fulvue process was announced in which anamorphic compression was used,[6] and a similar lens designed by Sidney H. Newcomer was made by the Goerz American Optical Company under the name of Cine-Panor for 16 mm movies. The idea was not developed commercially at that time, largely because of the Great Depression and the advent of sound movies. However, in 1952, following

---

[2] Ger. Pat. 99,722; Brit. Pat. 8,512/98.

[3] U.S. Pat. 1,032,172; Brit. Pat. 493/11; Ger. Pat. 232,848.

[4] U.S. Pat. 1,829,633/4; Brit. Pat. 349,905; Ger. Pat. 556,576.

[5] *J. Opt. Soc. Am.* **18,** 174 (1929).

[6] B. Coe, *The History of Movie Photography,* page 147. Eastview Editions, Westfield, N.J., 1981.

Figure 13.2. The Hypergonar lens.

the success of the Cinerama process, the Twentieth-Century Fox Film Corporation returned to the Chrétien principle and made their first Cine-maScope picture, "The Robe," using anamorphic camera and projector attachments.

Chrétien's original anamorphic attachment was called the Hypergonar (Fig. 13.2); it is described in his patent.[7] Large numbers of these lenses were manufactured by Bausch & Lomb for the CinemaScope process. When photographing a near object it is necessary to focus both the main lens and the attachment simultaneously, so for this and other reasons it is now common to photograph the original scene on wide film and compress the image to the usual 35 mm format in the printer when making release prints.

---

[7] U.S. Pat. 1,962,892; Brit. Pat. 356,955 (1929).

In his *Treatise on Optics* written in 1831, Sir David Brewster (1781–1868), explained how a pair of refracting prisms can be used to form an anamorphic telescope. If a prism is held before the eye at minimum deviation with the refracting edge vertical, the image is not distorted but it is deviated, and it appears colored because of the dispersion of the glass. If the prism is then tilted out of minimum deviation, the image becomes wider or narrower in lateral dimensions with no change in the vertical direction. By using two similar prisms in succession, one with its apex to the right and the other to the left, the chromatic dispersion and the deviation are eliminated and the anamorphic compression is doubled. Such an arrangement is known as a Brewster Prism (Fig. 13.3), and it can be used in front of a camera or projector as a continuously variable anamorphic attachment. It must be used in parallel light, however, and any small amount of chromatic dispersion at the ends of the field can be corrected by achromatizing the prisms. For use over a wide angular field or in nonparallel light, a system of cylindrical lenses is generally more satisfactory than a Brewster prism. Brewster pointed out that by using two pairs of prisms with their power meridians perpendicular, it is possible to construct a variable-power telescope having no anamorphic distortion. This instrument is known as the Teinoscope.

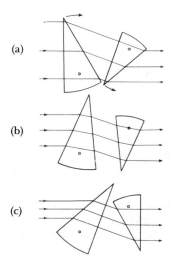

Figure 13.3. The Brewster prism anamorphoser: (a) mag. = 1.82; (b) mag. = unity; and (c) mag. = 0.55.

F. Deep-Field Lenses

It can be readily shown that the depth of field in a photograph depends mainly on the distance of the subject and the linear diameter of the lens aperture, while the exposure time depends only on the F-number. Consequently, the depth of field can be increased without loss of lens speed by going to a shorter focal length, and this is one of the reasons for the popularity of 35 mm cameras over those with a larger format. In fact, the depth of field of the current Kodak Disc camera, which is equipped with an $f/2.8$ lens of 12 mm focal length, is so great that no focusing means is required.

Before the miniaturization of everything in photography, opticians tried to find ways to increase the depth of field without having to stop the lens down to a small aperture. It had been suggested by Claudet as early as 1867 that the lens could be moved a short distance longitudinally during the exposure, so that there would be a sharp image of each object in the scene superposed on a somewhat blurred image of the same object.[8] Because the sharp image would be brighter than the blurred image it would be exposed more, provided that the exposure was on the lean side. If the picture were overexposed, of course, everthing would be recorded, with a hopeless amount of overall blur. Devices for accomplishing this movement were patented by Stark[9] and Nehring[10] in the early years of this century. It should be pointed out that the movement of the lens would lead to some degree of radial displacement of the outer parts of the picture. In 1927 Dieterich[11] applied this principle to a motion-picture camera by causing one element in the lens to be vibrated longitudinally during exposure, and he claimed that he had succeeded in maintaining the image size while the depth was increased. His novel lens was constructed by Bausch & Lomb under the name of Detrar.

Another way to increase the depth of field is by the deliberate introduction of axial aberration. In this way each object in the scene is imaged sharply by one zone of the lens or by light of one wavelength, while all the other zones and wavelengths produce superposed images, which are to some degree out of focus and therefore less likely to be exposed on the film. In 1928 Howard Beach of Buffalo, New York, offered for sale an

[8] *Phot. Journ.* **11,** 178 (1867).
[9] U.S. Pat. 751,116 (1903).
[10] U.S. Pat. 756,881 (1904).
[11] *Electronics* **15,** 44 (March 1942); U.S. Pat. 1,927,925 (1927).

attachment consisting of a plane glass plate having an aspheric surface polished on one side, which served to introduce some spherical aberration into the image.[12] He also made lenses of the Tessar type in which the front surface was hand polished to an aspheric shape; these so-called Multifocal lenses were sold by Wollensak.

A number of soft-focus portrait lenses have been manufactured in which a large amount of spherical aberration was deliberately introduced into the design. Among these are the Rodenstock Imagonal, the Busch Nikola Perscheid, and the lens by Pinkham and Smith of Boston. These lenses offered two advantages to the photographer: they exhibited an increased depth of field and the image had a type of fuzziness that was much admired in portraits. Further, by stopping the lens down, the amount of spherical aberration in the image could be varied as desired.

Another approach has been to introduce deliberately some degree of chromatic aberration into the image. In 1920 the Goerz Company announced the Mollar attachment for this purpose.[13] It consisted of a thick plane-parallel plate containing a buried surface (see Chapter 8, Section II). An example of pictures made with and without this attachment is shown in a book by Zeiss.[14] It is probable that this device would not be satisfactory if used for color photography.

II. INTERCHANGEABLE FRONT COMPONENTS

The first proposal to use a pair of interchangeable front components with a common rear component to provide two different lengths appears to have been the Tele-Ansatz introduced by Agfa in 1926.[15] The main lens was a 105 mm $f/6.3$ Cooke Triplet with the diaphragm located in the rear airspace. To increase the focal length to 210 mm, the front two elements were removed and replaced by a large cemented doublet followed at a considerable distance by a negative element, the F-number remaining unchanged by this substitution. It is not known how successful this system was in actual practice.

The next proposal was that used in the Retina IIIc camera manufac-

---

[12] U.S. Pat. 2,101,016; Brit. Pat. 335,696 (1929).

[13] Z. für Instkde. **44**, 310 (1924); Ger. Pat. 364,003; U.S. Pat. 1,556,982 (1920).

[14] W. Merté, R. Richter, and M. von Rohr, Das photographische Objektiv, page 234. Springer, Wien, 1932.

[15] Ger. Pat. 472,234 (1926).

tured by the Nagel Werk in Stuttgart in 1954. This system was based on a classical $f/2$ double Gauss lens in which the shutter, diaphragm, and rear lens component remained fixed in the camera. The front component was replaceable by either of two attachments, one to increase the focal length and the other to reduce it. Two systems were marketed, based on the Schneider Xenon and the Rodenstock Heligon.

To change the focal length of the Schneider lens from 48 mm to 76 mm, the front component was removed and replaced by the Longar Xenon (Fig. 13.4). This unit contained an afocal Galilean telescope followed by a cemented doublet, focusing on near objects being performed by varying the separation between the two parts of the Galilean telescope. The aperture of the new system was limited to $f/4$. For a wide-angle 35 mm lens, a different front component was used called the Curtar; this contained a reversed Galilean telescope followed by a cemented doublet; the aperture was now only $f/5.6$.

The design of these two converters must have presented a difficult problem. Eventually the whole system was abandoned in favor of a set of interchangeable complete objectives with a common shutter in the rear. Each of these objectives had its own iris diaphragm, but in order to provide reasonably uniform illumination across the frame, the rear element of

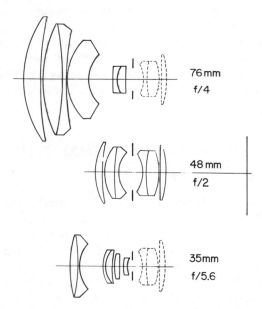

Figure 13.4. The Schneider convertible system for the Retina IIIc camera.

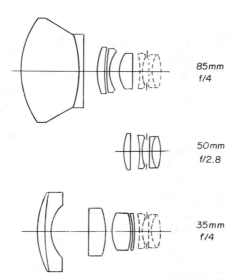

85mm
f/4

50mm
f/2.8

35mm
f/4

Figure 13.5. The Zeiss Satz-Tessar convertible system, with Pro-Tessar attachments.

each lens had to be located close to the small shutter opening. In spite of these difficulties, a series of acceptable objectives was produced with focal lengths ranging from 28 mm to 200 mm.

In 1957 Zeiss introduced a similar series of interchangeable front components for use with the rear component of a 50 mm $f/2.8$ Tessar (Fig. 13.5). The interchangeable fronts were known as Pro-Tessar lenses, giving focal lengths of 85 mm and 35 mm at $f/4$. These lenses were intended for use on the Contaflex camera.

III. REAR ATTACHMENTS

A. The Field Flattener

The use of a strong negative lens close to the focal plane of an ordinary objective has a powerful effect on the Petzval sum. This application was first suggested by C. Piazzi Smyth in 1873 (see Chapter 3, Section V). Such a lens also tends to overcorrect the astigmatism and bend the field backward, so that it must be designed specifically as a part of a complete objective system. In some aerial cameras the front surface of the glass platen, used to keep the film flat, is made concave in shape to act as a field flattener.

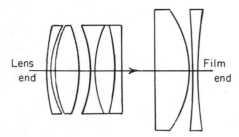

Figure 13.6. The Vivitar 2 × Macro Focusing Teleconverter.

B. Telenegative Attachment Lenses

The oldest type of rear attachment is a negative lens mounted at a suitable distance behind a normal camera objective to increase the focal length. This has already been discussed under Telephoto Lenses in Chapter 9. The advantage of this arrangement is that by varying the separation between the positive and negative components the overall focal length of the system can be varied in a continuous manner, thus constituting a primitive varifocal lens.

Of course, since the positive component is fully corrected in itself, the added negative component should be equally well corrected, but this is seldom the case. What actually happens is that the axial aberrations of the attachment are well corrected, but the oblique aberrations are corrected as well as possible for some assumed front objective at one specific position. When the separation is changed to vary the focal length, some loss of definition is bound to occur. Nevertheless, in spite of these difficulties, a number of 2 × and 3 × negative converters have been manufactured for SLR cameras, which work fairly well. The factor, 2 × or 3 ×, indicates the amount by which the focal length and the F-number of the main lens are changed when the converter is inserted between the main lens and the camera body. One convenient result of this arrangement is that the closest focusing distance of the main lens remains the same even though the focal length has been greatly increased.

Recently the Vivitar Company introduced a 2 × Macro Focusing Teleconverter (Fig. 13.6) in which a rapid-pitch focusing screw enables the main lens to be moved forward when focusing on a small object, so that a magnification as high as unity can be achieved. Of course, since the telenegative attachment doubles the image size, the main lens will be working at a magnification of 0.5 ×, which is greater than the lens was designed to cover. Consequently, it is advisable to stop the system down as far as possible when working at such high magnification ratios.

CHAPTER 14

Brief Biographies

In this chapter I am giving photographs and brief biographies of a few of the more notable opticians and lens designers of the past. This has proved to represent a difficult search. In some cases I have found the life story of an individual but no portrait, while in other cases I have a portrait but have been unable to find out anything about that person's history.

It is recognized that this is only a very incomplete list of the hundreds of designers and manufacturers who have contributed to the development of optical glass and lenses of all types over the years. I have naturally concentrated on photographic lenses, but not exclusively as will be seen. References to further information are given to English-language publications wherever possible.

If anyone can add information or portraits to this collection it will be gratefully received.

List of Biographies in Order of Date of Birth

1732	J. C. Voigtländer	1779	J. F. Voigtländer
1748	P. L. Guinand	1787	J. Fraunhofer
1766	W. H. Wollaston	1794	G. S. Plössl

1798	A. Ross

1800	I. Holmes
1800	T. Grubb
1801	C. A. Steinheil
1804	C. L. Chevalier
1807	J. M. Petzval
(1810?)	J. T. Jamin
1812	P. W. F. Voigtländer
1818	T. Ross
1819	C. Piazzi Smyth
1819	T. Sutton
1820	E. Busch
(1820?)	C. C. Harrison
1826	J. Zentmayer
1828	A. Darlot
1830	J. H. Dallmeyer
1832	A. G. Clark
1832	H. A. Steinheil
1834	H. L. H. Schroeder
1834	E. W. Gundlach
1836	R. Morrison
1840	E. C. Abbe
1844	H. Grubb
1846	F. R. Voigtländer
1851	F. O. Schott
1854	E. Bausch
1856	C. P. Goerz
1858	C. Moser
1858	P. Rudolph
1859	T. R. Dallmeyer

(1860?)	J. Schneider
1862	H. D. Taylor
1862	A. Wollensak
1864	J. C. Wollensak
1865	E. von Höegh
1865	R. Steinheil
1866	A. E. Conrady
1868	H. Harting
1868	L. O. M. von Rohr
1870	W. O. W. Zschokke
1870	C. W. Frederick
1873	G. A. H. Kellner
1879	B. V. Schmidt
1879	H. J. Chrétien
1879	E. Wandersleb
1886	R. Richter
1886	M. Berek
1889	W. E. Schade
1889	W. Merté
1889	H. W. Lee
1892	J. B. Walker
1891	A. Warmisham
1893	F. E. Altman
1893	A. Bouwers
1896	D. D. Maksutov
1900	L. J. Bertele
1902	F. G. Back
1903	R. H. Naumann
1907	P. Angénieux

1869

1893

Ernst Carl Abbe

1840 – 1905

Ernst Carl Abbe was the son of a foreman at a textile factor in Eisenach. He attended the universities of Jena and Göttingen, where he obtained a PhD degree. After a brief period as a tutor, he became an instructor in physics and mathematics at Jena at the age of 23. Soon he was sought out by Carl Zeiss, who needed someone to help him with his attempts to design a microscope; eventually Abbe decided to leave the university and work full-time with Zeiss. He married in 1871 and had two daughters.

While at Zeiss, Abbe worked mainly on the microscope; he contributed greatly to the theory of that instrument. In 1883 he hired Otto Schott to help develop some new types of glass that he needed to reduce the secondary spectrum of microscope objectives. Together they founded the Schott Glassworks in Jena, where many new types of optical glass were developed and offered for sale. Abbe then hired Dr. Paul Rudolph to design photographic objectives using the new types of glass.

On Carl Zeiss's death in 1888 he bequeathed the company jointly to his son Roderich and Abbe, but they could not agree on policies, so Abbe bought out Roderich's share and became sole owner of the company. Abbe was a great humanitarian and he soon handed over the company to the Carl Zeiss Foundation, of which the workers were part owners. He introduced many unheard-of reforms such as the eight-hour day, sick benefits, and paid vacations. He died in 1905, largely as the result of overwork.

REFERENCES: F. Auerbach, *The Zeiss Works,* trans. R. Kanthack. Foyle, London, 1927; H. Volkmann, "E. Abbe and his Work," *Appl. Opt.* **5,** 1720 (1966); *Dictionary of Scientific Biography* **1,** 6 (1970).

1960

Fred Early Altman

1893 – 1964

Fred Altman was born in Altoona, Pennsylvania, and his family moved to Des Moines, Iowa, where he grew up. He had attended Iowa State College for about a year when his family circumstances made it necessary for him to go to work. He entered the Kodak Research Laboratories in the fall of 1915 at the age of 22 as a technician in the physics department. After six months he was transferred to the newly established Scientific Department at Hawk-Eye Works as assistant to C. W. Frederick. Altman proved to be an exceedingly apt pupil with a genius for intuitive lens design. Within a few months working with logarithms, he was designing excellent lenses, including a four-element anastigmat that was manufactured extensively. During World War II Altman worked on several large lenses for aerial cameras. He continued to be responsible for many of Kodak's lenses until his retirement in 1960 at the age of 67. He died in Rochester four years later.

1957

Pierre Angénieux

1907 –

Pierre Angénieux was born in July 1907 at St. Héand, Loire, where he later established his lens factory. He attended high school in St. Etienne and engineering college in Cluny from 1925 to 1928. He then did two years of graduate work at the École Supérieur d'Optique in Paris. He married in 1934 and had four children.

His industrial career included two years with Pathé Cinema in Paris from 1930 to 1932, followed by two years as chief engineer for cine lenses at OPTIS in Paris. He then became a partner at ASIOM from 1933 to 1935, working on lenses for additive color movies. In 1935 he established his own company in Paris with a factory at St. Héand. There he developed a series of reversed-telephoto lenses with the name Retrofocus; this is now becoming a generic term for all lenses of this type. Since 1957 he has developed many zoom lenses, the first of which had a 4:1 focal-length range; some 70,000 were manufactured. In 1961 he extended his designs to include an 8:1 zoom for 8 mm movie cameras and a 10:1 zoom for 16 mm equipment and television cameras. Employees at his factory numbered eighty in 1945, but the number quickly rose to nearly one thousand in later years.

Mr. Angénieux has received many honors for his achievements, including the French Legion of Honor and an Oscar from the Society of Motion Picture Arts and Sciences. He retired in 1974 at the age of 67 and moved to Geneva, Switzerland. The company is now in the hands of his eldest son, Bernard.

REFERENCES: *Who's Who in France, 1984–1985;* Also, a letter from his son Bernard.

Frank G. Back

1902 – 1983

Frank G. Back was born in Vienna and obtained the degrees of ME in 1925 and ScD in 1931 from the University of Vienna. He came to the United States in 1939 after serving as a consulting engineer in Vienna and Paris.

During World War II he developed a zoom viewfinder for the armed forces, and after the war he established the Viewfinder Corporation in New York City, where he designed the first Zoomar lens for use on 16 mm and 35 mm motion-picture cameras. In 1951 he established the Zoomar Corporation on Long Island, where numerous zoom and other lenses were manufactured. He designed the first zoom lens for a still camera in 1958. He moved to LaJolla in 1981, where he died two years later. Dr. Back received many awards, including the Progress Medal of the Society of Motion Picture and Television Engineers (SMPTE) in 1962.

REFERENCE: *JSMPTE* **92**, 1105 (October 1983).

1890

1933

Edward Bausch

1854 – 1944

Edward Bausch was the eldest son of J. J. Bausch, founder of the Bausch & Lomb Optical Company. He was born in Rochester in 1854 and attended Cornell University as an engineering student. On his graduation in 1874 he urged his father to expand the company's activities beyond the making of spectacle lenses; his first choice was the microscope, for which there was a considerable demand at that time. With the help of Ernst Gundlach the company started to manufacture microscopes, which were very successful. They added photographic lenses to their line in 1883, mainly Zeiss designs made under license. Edward Bausch designed the Iris Diaphragm shutter in 1888 and the Plastigmat lens in 1900. By 1903 the firm had made about 500,000 camera lenses and a like number of shutters.

Edward Bausch wrote two books on the microscope, one in 1885 and the other in 1898. He became president of the company on the death of his father in 1928 and Chairman of the Board in 1935. He was awarded an honorary doctorate by the University of Rochester in 1931. With his brother William he started to manufacture optical glass in 1915 when all supplies from Europe were cut off by World War I. He was a great philanthropist and presented the city with a large new building for their museum in 1942. He received many honors during his long life of just under ninety years.

REFERENCE: E. R. Foreman, *Edward Bausch.* Society of the Genesee, 1933.

Max Berek

1886 – 1949

Max Berek was born at Racibórz in South Poland, near Katowice, in the ore and coal district of Oberschlesia. He was attracted early to the mysteries of crystalline materials and after graduation from the local gymnasium he obtained practical experience at the Königshütte in Oberschlesia. He then became a science student at Berlin University, concentrating on mineralogy. In 1910 he was appointed assistant in the Mineralogical Institute in Berlin, working under T. Liebisch, where he obtained a PhD degree in crystal optics.

In March 1912 he joined the firm of Ernst Leitz in Wetzlar, where he remained until his death 37 years later. His scientific work was devoted almost entirely to the theory of the microscope, particularly the polarizing microscope used in mineralogy. When the Leica camera was being developed by Oskar Barnack in the early 1920s, Berek designed several suitable lenses, including the Elmar, Summar, Summitar, and Hektor series. He wrote a book on practical optics and over eighty scientific papers.

REFERENCES: "Nachruf auf Prof. Dr. Phil, h. c. Max Berek." *Optik* **6,** 311 (1950); *Leitz Mitteilung* **9,** 1 (January 1986).

1923

ca 1950

Ludwig Jakob Bertele

1900 – 1985

Ludwig Jakob Bertele was born in Munich in 1900, the son of an architect. He attended the Institute of Technology in Dresden and in 1920 became an optical designer with Ernemann in that city. There, with no formal optical training, he designed the first $f/2$ Ernostar lens in 1922. This lens was fitted to the Ermanox camera, and its use permitted the first photographs to be made under available light.

In 1926 Ernemann became part of the Zeiss-Ikon organization and Bertele became a Zeiss designer, working first at Jena and then at the ICA plant in Dresden, where he designed a series of high-aperture Sonnar lenses. He quickly became one of the most outstanding lens designers in the world.

In 1942 Bertele left Zeiss and worked for three years with Steinheil in Munich. Then, at the close of the war in 1945, he moved to Switzerland and joined the firm of Wild in Heerbrugg. There he designed microscope objectives and a series of outstanding lenses for aerial cameras. He was awarded an honorary doctorate by ETH in Zurich in 1959. Bertele retired to Wildhaus in 1956 but continued his optical work on a consulting basis until his retirement in 1973. He received many honors for his outstanding accomplishments; he died on November 16, 1985.

REFERENCES: *Festschrift Dr. Bertele,* published by Wild in 1975; *Who's Who in Switzerland, 1984–85.*

Albert Bouwers

1893 – 1972

Albert Bouwers was born at Dalen in the Netherlands. He attended the universities of Amsterdam and Utrecht, and joined the Philips Research Laboratories in 1920 where he eventually became head of the high-voltage and X ray research division. In 1941 he patented his concentric mirror-lens, which he developed as an aid in X ray fluorography. At the end of World War II, in 1946, he published his well-known book *Achievements in Optics,* in which he described many other concentric mirror systems.

His interest in optics was so great that in 1951 he left Philips and founded the Old Delft Optical Company, from which he retired in 1968; he died four years later. This company is still in existance, making a variety of large lens systems for aerial and other photographic applications. Bouwers was a prolific writer, with twenty-four books and over one hundred papers to his credit. He received an honorary doctorate from Delft in 1963.

REFERENCE: *Albert Bouwers Selected Scientific Papers,* North Holland, 1969.

Emil Busch

1820 – 1888

It appears that there was a spectacle factory in Rathenow founded in 1800 by a preacher named Johann Duncker. On his death in 1824 he was succeeded by his son Edouard and, as the son had no children, the works was taken over by Duncker's nephew Emil Busch in 1845. Busch immediately enlarged the factory, installed a steam engine, and began to make photographic objectives, notably several Petzval lenses of six or seven inches in aperture. In 1865 Busch patented the Pantoskop lens, the first true anastigmat, which was made for several years as an excellent wide-angle objective, somewhat better in performance than the famous Globe lens of Harrison and Schnitzer. Later the company became known as the ROJA (Rathenower Optische Institute) and eventually Emil Busch AG. The Busch Company is still in existence making all types of ophthalmic machinery.

REFERENCE: Emil Busch *Hausmitteilungen* **16,** 13 (March – May 1938).

Charles Louis Chevalier

1804 – 1859

Charles Chevalier was the son and grandson of Paris opticians, and at the age of 17 joined his father in the manufacture of microscopes and other optical devices. In 1832 he left the family home on the Quai de L'Horloge and settled in the courtyard of the Palais Royale. He was a vigorous worker and wrote several books on the microscope, spectacles, and photography. His invention of a new type of microscope objective earned him two gold medals in 1834.

Chevalier made the first lenses used by Daguerre in his photographic experiments; the Giroux whole-plate camera of 1839 was equipped with a Chevalier landscape lens. He attempted to make a portrait lens in 1840, but it was not successful. He died in 1859 at the age of 55 and was succeeded by his son Louis-Marie Arthur, who wrote a biography of his father in 1862.

1956

Henri Jacques Chrétien

1879 – 1956

Henri Chrétien was born in Paris, the son of a merchant. He married in 1910 and had three children. He attended a professional school of typography and the Superior School of Electricity, where he obtained a bachelor's degree in science. He then worked at Meudon Observatory and in 1910 became Chief of Astrophysics at Nice. He helped to establish the Institut d'Optique in Paris early in 1917. He was professor of astrophysics at the Sorbonne in 1927, where he demonstrated the use of an anamorphoser for wide-screen motion pictures. This system was shown in Washington in November 1928. In 1952 Twentieth-Century Fox secured rights under Chrétien's patent, which became the basis of their CinemaScope process.

Chrétien collaborated with G. W. Ritchey in the development of the Ritchey-Chrétien telescope, a 40-inch version being installed at the Naval Observatory in Washington in 1932. He wrote a large book entitled *Cours de Calcul des Combinaisons Optiques,* which went through several editions. He was living in Washington at the time of his death in 1956.

REFERENCES: *JSMPTE* **65,** 110 (1956); *J. Opt. Soc. Am.* **18,** 174 (1929); *Who's Who in France, 1955 – 56.*

Alvan Graham Clark

1832 – 1897

The Alvan Clark Company of Cambridge, Massachusetts, was founded in 1850 by two brothers, George Bassett Clark (1827 – 1891) and Alvan Graham Clark. Their father, Alvan Clark Sr. (1804 – 1887) was a portrait painter, but he was interested in mechanical things, and in 1860 he closed his Boston studio to devote his whole time to making telescopes. The company made a large number of instruments that were greatly admired, culminating in the Lick 36-inch of 1888 and the Yerkes 40-inch of 1897.

Alvan G. Clark's only foray into photographic optics was in an 1888 patent, in which he suggested that a pair of Gauss-type meniscus lenses could be mounted back-to-back in a tube to form a photographic objective. Actually, the Clarks had made only one Gauss-type objective, a 9-inch for Princeton in 1877, and they considered the gain too slight to be significant. The Alvan G. Clark Lens was included in Bausch & Lomb's catalogs from 1890 to 1898, but apparently very few were sold. However, the idea was sound, and it led eventually to the well-known double-Gauss lenses of today.

REFERENCE: D. J. Warner, *Alvan Clark and Sons.* Smithsonian Press, Washington, D.C., 1968.

1930

Alexander Eugen Conrady

1866 – 1944

Alexander E. Conrady was born at Burscheid, near Bonn, in January 1866, the fifth of a family of seven children. His father, grandfather, and great-grandfather had been schoolteachers, but his father later decided to go into business. Young Alexander was a studious child and in 1884 went to Bonn University, where he studied chemistry, physics, mathematics, and astronomy. In 1886 his father became German agent for an English company, Stansfield, Brown and Co., and young Alexander went to England twice on his father's behalf. He finally decided to live in England permanently and traveled to America and South Africa for Mr. Brown. He then set up a small factory in Leeds to make electrical machinery. In 1900 he moved to London, where he established a workshop making microscope objectives, which were sold by Watson's. In 1902 he joined Watson's staff, where he remained until 1917, when he was appointed professor of optical design at the newly established Technical Optics Department at the Imperial College in London. He published his well-known book *Applied Optics and Optical Design* in 1929 and retired two years later at age 65. He died during the war, in June 1944.

REFERENCE: *Applied Optics* **5,** 176 (January 1966).

John Henry Dallmeyer

1830 – 1883

John H. Dallmeyer was born in Loxten, Westphalia, and apprenticed to an optician in Osnabrück. He emigrated to England in 1851 and after working at several small jobs entered the firm of Andrew Ross (1798 – 1859). He was highly respected by Ross and eventually married Ross's second daughter Hannah. When Ross died he left one-third of his considerable fortune — estimated at over £ 20,000 — to Dallmeyer. The following year Dallmeyer and Ross's son Thomas (1818 – 1870) agreed to separate, and Dallmeyer established his own business in London. He worked diligently on the improvement of photographic lenses, introducing in succession various high-aperture portrait lenses, and in 1862 he produced his Triple Achromatic lens. In 1866 he introduced his Patent Portrait lens, the Wide-angle Rectilinear, and, finally, his most important invention, the Rapid Rectilinear.

Dallmeyer received many honors, including the French Legion of Honor and the Russian Order of St. Stanislaus. He died in December 1883 at the age of 53, while on a long sea voyage for his health. He left a second wife and five children. His eldest son, Thomas, took over management of the company.

REFERENCE: *Dictionary of National Biography,* Vol. **V**, p. 400. Oxford, 1917.

ca 1904

Thomas Rudolph Dallmeyer

1859 – 1906

Thomas R. Dallmeyer was the eldest son of J. H. Dallmeyer. He attended Mill Hill School near London and studied optics at Kings College, London, under Sylvanus Thompson, where he obtained a BSc degree. He became manager of the company some time before his father's death in 1883. In 1891 he became fascinated with the variable-power telephoto lens and felt that there was some magic about it; indeed, he wrote *Telephotography,* a book dealing with this system and what could be done with it. In 1893 he worked with J. S. Bergheim on the development of an uncorrected tele-photo lens for soft-focus portraiture. He received the Progress Medal of the Royal Photographic Society in 1896 and became president of the Society from 1900 until 1902. He died in December 1906 at the age of 47.

In 1892 the firm became a Limited Company under the chairmanship of James Ludovic Lindsay, 26th Earl of Crawford; on his death in 1912 management fell to C. F. Lan-Davis. The company is still in existence in London.

REFERENCE: *Phot. J.* **47,** frontispiece (January 1907).

Mr A. Darlot

Alphonse Darlot

1828 – 1895

Alphonse Darlot began his optical training by being apprenticed to Lerebours and Secrétan in Paris. He later joined J. T. Jamin, whom he succeeded as manager of the factory in 1860 at age 32. Their company was then at 14 rue Chapon, but Darlot moved it to 125 Blvd Voltaire in 1877. He was succeeded on his death in 1895 by L. Turillon.

Darlot made a large number of photographic lenses, mainly of the Petzval type, but he also made many landscape lenses and Rectilinears under the name of Hemisphérique and Hemisphérique Rapide. Some of his small lenses had three swing-out stops, which were characteristic. He also manufactured cameras and other photographic accessories. He was awarded a silver medal by the Paris International Exhibition of 1867 and was made a Knight of the Legion of Honor in 1892.

REFERENCE: *Image,* George Eastman House, Rochester **11**, 22 (1962).

Joseph Fraunhofer

1787 – 1826

Joseph Fraunhofer was born at Straubing, Bavaria, the eleventh and last child of Xavier Fraunhofer, a worker in decorative glass. His parents died when he was only twelve, and his guardian apprenticed him to a mirror maker named Weichsberger. In 1806, at age 19, he was offered a position in the Munich Mathematical-Mechanical Institute, which had been established three years previously by Reichenbach, Utzschneider, and Liebherr. Soon he was grinding lenses by himself, and he was placed in charge of the glass operations at the old monastery at Benediktbeuern. He soon quareled with the glassmaker Guinand, who left the company and returned to Switzerland in 1814. Fraunhofer had meanwhile become a junior partner in the firm and was made director in 1820. He made several large telescopes, culminating in the 9½-inch telescope at Dorpat, now Tartu, Estonia, in 1824. He worked on many other problems, including the making of diffraction gratings. He died in 1826 at the age of 39.

REFERENCES: M. von Rohr, *Joseph Fraunhofers Leben, Leistungen und Wirksamkeit, Akademische Verlagsgesellschaft,* Leipzig, 1929; M. von Rohr, "Fraunhofer's Work and its Present-day Significance." *Trans. Opt. Soc. (London)* **27,** 277 (1926).

Charles Warnock Frederick

1870 – 1942

Charles W. Frederick was born in Des Moines, Iowa, and after graduation from the University of Kansas in 1892 decided to become a civil engineer. In 1901 he took a Civil Service examination and secured an appointment at the Naval Observatory in Washington. Between 1904 and 1906 he was sent by the Observatory to Samoa to observe southern stars. In 1909 he was appointed professor of mathematics at the Naval Academy in Annapolis.

In 1914 Mr. Frederick was invited to establish a lens-design department at the Eastman Kodak Company in Rochester New York, and although he knew nothing about lens design, he accepted the invitation. He was aided in his task by F. E. Ross, G. W. Moffitt, G. S. Dey, and Max Zwillinger. In 1916 he was joined by F. E. Altman, their first job being to design aerial camera lenses for the armed forces. After World War II they went on to design a large number of lenses for Kodak equipment. Frederick retired in 1938, to be succeeded by R. Kingslake, and died four years later.

REFERENCE: *Appl. Opt.* **11,** 50 (1972).

Carl Paul Goerz

1856–1923

Carl P. Goerz was born in Brandenburg and after leaving school was apprenticed to the firm of Emil Busch. In 1886 he took over a small instrument merchandising business in Berlin-Schöneberg to supply schools. The following year he added photographic supplies to his line and, with the help of Carl Moser (1858–1892), began the manufacture of photographic lenses the following year. On Moser's death in 1892, Goerz hired Emil von Höegh (1865–1915), who brought to the company the Dagor lens, of which some 30,000 were sold in the first four years. Von Höegh resigned in 1902 for reasons of ill health and was succeeded by the Swiss designer Walter Zschokke. Other designers were F. Urban and, briefly, Robert Richter.

Goerz set up a small lens factory at Winterstein in 1895 and a larger one at Friedenau in 1898, where a wide range of lenses was made. He added shutters in 1896 and prism binoculars in 1899. In 1910 he established a film factory in Berlin. The growth of the company was little short of fantastic. With 25 employees in 1890 the number grew to 200 in 1895, 1,000 in 1900, and 3,000 in 1914. During the first World War Goerz manufactured large numbers of optical devices for the armed services. In 1921 Goerz cooperated with Steinheil in the establishment of the Sendlicher Optical Glass works in Berlin. After the war, only Zeiss was permitted to manufacture military equipment, and Goerz's business declined badly. Carl Goerz died in 1923; three years later the company was acquired by the Zeiss-Ikon combine, although the branches in Vienna and New York continued under their old names.

REFERENCES: J. M. Eder, *History of Photography,* trans. E. Epstean, p. 409. Dover, New York, 1978. F. Wentzel, *Memoirs of a Photochemist,* pages 3 and 122. American Museum of Photography, Philadelphia, 1960.

Thomas

Sir Howard

Thomas Grubb

1800 – 1878

Sir Howard Grubb

1844 – 1931

Thomas Grubb was born at Kilkenny in Ireland and was always interested in mechanical things. In 1840, while living in Dublin, he was appointed engineer to the Bank of Ireland, where he designed and built machinery for the printing and numbering of bank notes. In 1834 Grubb became interested in optics and set up a workshop to make telescopes, his greatest work being the 48-inch Melbourn reflector of 1862. He designed his Aplanat camera lens in 1857, several hundreds of which were sold over the following years. In 1864 he was made a Fellow of the Royal Society (FRS).

Grubb's son Howard was born in Dublin and studied civil engineering at Trinity College. At the age of 20 he joined his father in the telescope factory, which was located first at Charlemont Bridge and later at Rathmines. In 1880 he built a 27-inch refracting telescope for Vienna, which was for several years the largest refractor in the world. This was followed by the Greenwich 28-inch in 1893. Grubb was made an FRS in 1883 and was knighted in 1887.

In 1918 their telescope factory was moved to St. Albans in England so as to be closer to the war effort. There they continued to make astronomical instruments until 1925, when the company was taken over by Sir Charles Parsons and moved to Walkergate near Newcastle. The Grubb-Parsons Company ceased operations in 1984.

REFERENCES: *Parsons Journal* **11,** 145 (1967); *Sky and Telescope* **67,** 227 (1984).

Pierre Louis Guinand

1748–1824

Pierre L. Guinand was born at La Corbatière in Switzerland. He was a maker of clock cases and became interested in lens manufacture about 1783. Finding that flint disks were filled with striae, he studied glass chemistry and in 1805, at the age of 57, discovered the process of stirring the melt with a ceramic stirrer to remove the striae. His flint disks were so fine that Utzschneider invited him to join his Institute and establish a glass factory at Benediktbeuern in Bavaria in 1806. After Fraunhofer's arrival the two men worked together from 1811 to 1813, improving their glass enormously. However, they eventually quarreled, and Guinand returned to Switzerland and reopened his glassworks at Les Brenets in 1818, with the help of his wife Rosalie and his son Aimé. On Guinand's death in 1824 at the age of 76, Aimé let the works deteriorate, but it was taken over by his oldest son, Henri, who took it to France where it eventually led to Parra-Mantois and Chance in England.

REFERENCE: Parra-Mantois catalog; *Trans. Opt. Soc.* **27**, 277 (1926); *Z. für Instkde* **46**, 121 (1926).

1896

Ernst W. Gundlach

1834 - ?

Ernst W. Gundlach was born in Pyritz, East Prussia, and was apprenticed to Lewert's workshop in Berlin. He then worked at various optical factories in Vienna, Paris, and London, trying to find out how to make lenses but always frustrated by the jealousy of the workers there. In 1859 he was employed by C. F. Belthle in Wetzlar, a firm that eventually became Leitz, and in 1864 he returned to Berlin where he established a small workshop to make microscope objectives.

In 1872 he sold out to Seibert and Krafft and emigrated to the United States, where he settled in Hackensack, New Jersey, making microscope objectives, many of which were much admired by microscopists. In 1876 he was hired by Bausch & Lomb in Rochester, New York, to assist in starting a microscope factory. He left Bausch & Lomb in 1878 to set up his own works, first in Rochester and then in Hartford, Connecticut. He returned to Rochester in 1884, where he established the Gundlach Optical Company to make microscopes and later photographic objectives. He patented the Rectigraphic lens in 1890. In 1895 he separated from the company and established a rival firm with his son, also making photographic objectives, but they were unsuccessful, so in 1898 they moved to Chicago where he joined the Vive Camera Company. He returned to Germany in 1899. His date and place of death are unknown.

REFERENCE: R. Kingslake, "Ernst Gundlach: Nineteenth Century Pioneer Optician," *History of Photography* **2,** 361 (1978).

1852

Charles C. Harrison

d. 1864

Charles C. Harrison first appeared in the New York City directories of 1847 as a daguerreotypist in City Hall Place and later on Broadway. In 1854 when the daguerreotype was losing favor Harrison sold his studio to George S. Cook, one of Brady's cameramen, while Harrison learned the art of lens making from Henry Fitz.

In 1860 Harrison reappeared in New York as a maker of cameras and lenses. In that year he patented the Globe lens with J. Schnitzer, and in 1861 he patented a novel iris diaphragm with both studs at the same end of the leaves. This type of diaphragm is universally employed today in cameras having automatic aperture control. In May 1863 it was stated that Harrison had made 8,800 ordinary lenses and 307 Globes. On Harrison's death in 1864 his factory was taken over by an employee named Richard Morrison (1836–1888).

REFERENCES: W. Welling, *Photography in America*, p. 108 T.Y. Crowell, New York, 1978; M. A. Root, *The Camera and the Pencil*, Philadelphia, 1864, p. 375.

Hans Harting

1868 – 1951

Hans Harting was born at Rummelsburg, near Berlin, and studied mathematics, physics, and astronomy at Berlin and Munich, where he obtained a PhD in 1889. For four years he was assistant to Auwers at the Prussian Academy of Sciences in Berlin and then transferred to the Physikalisch Technischen Reichsanstalt in 1893. He joined Zeiss in 1897, where for two years he was personal assistant to Ernst Abbe. For some reason he left Zeiss in 1899 to join Voigtländer in Brunswick, where he designed several well-known photographic objectives such as the Heliar and Dynar. He then joined the German Patent Office from 1908 to 1934, becoming department manager in 1922 and president in 1933. In 1934 he returned to Zeiss as a member of the Management Committee and retired in 1940 at the age of 72.

After World War II he was recalled to VEB Carl Zeiss in Jena to help with the postwar reconstruction. In 1950 he was made an Honorary Member of the Academy of Sciences in Berlin. Harting died at the age of 83 in 1951.

REFERENCES: *Jena Review, English Edn.* **13,** 250 (1968); also, H. Schrade, *Jenaer Jahrbuch for 1951.* Frontispiece and biography.

Emil von Höegh

1865 – 1915

Emil von Höegh was a descendent from an old Danish aristocratic family. He was born in Löwenberg, Silesia, and worked in Carl Bamberg's instrument workshop in Berlin. He also worked with Hartmann and Braun in Bockenheim and joined Paul Wächter in Friedenau in 1891.

Working privately, he developed a new type of photographic objective, and in the late summer of 1892 he applied for a position with Goerz after having been rejected by Zeiss. He presented Goerz with the design of a six-element symmetrical double anastigmat, a sample of which was made up and found to be excellent. It was immediately patented and put on the market under the name of Double Anastigmat Goerz, which was changed in 1904 to the acronym Dagor. Von Höegh was hired to replace Goerz's designer Moser, who had recently died, and he went on to design a four-element objective, which was called at first the Double Anastigmat Goerz Type B, the name being later changed to the Celor. He also designed the wide-angle Hypergon in 1900. Von Höegh resigned in 1902 for reasons of ill health; he moved to Rostock and later to Goslar, where he died twelve years later.

REFERENCE: R. Schwalberg, *Pop. Phot* **70,** 56 (January 1972).

Israel Holmes

1800 – 1874

The firm of Holmes, Booth and Haydens was incorporated in 1853 in Waterbury, Connecticut, to make brass hardware items, including those for the photographic trade such as silvered daguerreotype plates. The founder and president of the company was Israel Holmes; he had previously worked for Scovill and had been president of the Waterbury Brass Company. He continued in office until 1869, when he retired to establish another company. The other officers were John C. Booth (1808 – 1886) and two cousins, Henry W. Hayden and Hiram W. Hayden, both of whom were born in 1820.

Hiram Hayden was a keen daguerreotypist, and it was undoubtedly he who induced the company to make portrait lenses. A good many of these are still in existence, although some users considered them inferior to those of Harrison.

REFERENCE: J. Anderson, *The Town and City of Waterbury, Connecticut,* Vol. **II,** p. 352. Price and Lee, 1896.

Jean Théodore Jamin

d. 1867

The well-known firm of Jamin, opticien, Paris was established in 1822 by J. T. Jamin, and in 1849 it was located at 71 rue St. Martin in Paris, near Nôtre Dame. In 1850 it was moved to No. 127 in the same street and in 1856 to 14 rue Chapon nearby. Jamin himself continued to live at No. 127 until his death eleven years later.

Jamin made a large number of photographic objectives, mostly of the Petzval type, including many with a conical housing known as the Cône Centralisateur objective. The purpose of the conical housing was to prevent internal reflections from reaching the photographic plate. In 1860 Jamin retired and handed over management of the company to his employee Alphonse Darlot. For a year their lenses were labeled Jamin and Darlot, then the name Jamin was dropped.

G. A. Hermann Kellner

1873 – 1926

Dr. G. A. Hermann Kellner was born in Germany in July 1873, and studied at the universities of Berlin and Jena, where he received a PhD degree in 1899. After working for several years in the German optical industry, he came to the United States and was appointed director of the Scientific Bureau at Bausch & Lomb in Rochester. There he worked mainly on the microscope and fire-control instruments and later on projection systems.

He was a charter member of the Optical Society of America and the first editor of its journal (from 1917 to 1919).

REFERENCES: *J. Opt. Soc. Am.* **12,** 611 (1926); *SMPE,* **Trans #24,** page 162.

1961

Horace William Lee

1889 – 1976?

Horace W. Lee, one of England's foremost and most original lens designers, was born in January 1889 and obtained a BA degree at Cambridge University in 1911. He worked as an optical designer at Taylor, Taylor and Hobson, Leicester, from 1913 to 1936, where he designed a series of remarkable photographic objectives. His most outstanding designs were the Opic of 1920, the $f/2$ Speed Panchro, Hollywood's most used camera lens, in the late 1920s, and the reversed telephoto lenses used on the three-strip Technicolor camera.

In 1936 Lee moved to London and joined Scophony, where he remained for some ten years. He then joined Pullin and after a while transferred to Aldis in Birmingham. Prior to 1945 he wrote many articles in scientific journals on lenses and other optical subjects. He died sometime in the late 1970s.

Dmitri D. Maksutov

1896 – 1964

Dmitri D. Maksutov was born in April 1896 and graduated in 1913 from the Odessa Cadet Corps and in 1914 from the Military Engineering College. He worked first at Pulkovo Observatory until 1930, when he became director of the Laboratory of Astronomical Optics at Leningrad. In 1944 he was appointed professor at the State Optical Institute in Moscow. He returned to Pulkovo as associate in 1952, working mainly on methods for the manufacture and testing of large optical equipment. His many books included *Reflecting Surfaces and Systems* (1932); *The Shadow Method of Testing* (1934); *Astronomical Optics* (1946); and *The Manufacture and Testing of Astronomical Optical Systems* in (1948). In 1944 he became known in this country through his paper, "New Catadioptric Meniscus Systems," in *J. Opt. Soc. Am.* **34,** 270 (1944). He received many honors and awards, including the Badge of Honor and two Orders of Lenin. He died in August 1964.

REFERENCE: *Who Was Who in the USSR.* Scarecrow Press, Metuchen, N.J., 1972.

Willy Merté

1889 – 1948

Willy Merté was born in Dresden in January 1889, but he spent most of his youth in Weimar. He studied mathematics and physics at Munich and Jena, where his instructors included Röntgen and Wien. Merté obtained a doctorate in 1912 and passed his state examination for an instructor in 1913. Instead, he entered the photographic department at Zeiss, at the age of 24. There he designed many of their best lenses, including the Biotessar, Orthometar, Biotar, and Sphärogon. Many of the lens patents in the Zeiss name were actually designed by Merté. He was wounded in the lung during World War I.

After World War II, in 1945, he, along with other Zeiss scientists, left Jena and moved to the new establishment at Oberkochen. In 1947 he came to the United States and died in Dayton after a short illness in May 1948.

REFERENCE: *Optik* **7**, 121 (1950).

Richard Morrison

1836 – 1888

Richard Morrison was born in Gloucester, England, and at the age of 14 was apprenticed to Slater, a maker of astronomical lenses. He came to the United States in 1858, at the age of 22, and worked with Benjamin Pike, maker of telescope and microscope lenses. About 1860 he became foreman at the Harrison optical works and, on Harrison's death in 1864, he took over management of the firm in partnership with George Wale. Shortly thereafter he returned briefly to Benjamin Pike.

The New York Optical Works then sought his assistance, so he left Pike to join them. On the death of the manager, Mr. Schnitzer, Morrison determined to set up for himself, along with his former associate George Wale. He developed a number of unusual lenses, including the much admired Wide-angle lens, his whole output being taken by the Scovill Manufacturing Company. Morrison died in 1888 at the age of 52 after a three-year battle with tuberculosis.

REFERENCE: *Photographic Times and American Photographer* **18,** 578 (December 7, 1888).

Carl Moser

1858 – 1892

Carl Moser was born in Berlin and was hired by H. L. H. Schroeder as a tutor to his children. He was eventually taken into partnership with Schroeder and introduced to optical calculations. In the late 1870s Schroeder moved from Hamburg to Oberursel near Frankfurt, taking Moser with him. In 1881 Moser returned to Berlin, working on the design of telescopes with the optician O. Derge. After a brief return to Oberursel, where Schroeder's workshop had become bankrupt and Schroeder himself had to move to London, Moser returned once more to Berlin at the urging of Carl Bamberg and worked mainly on the design of telescope objectives.

In 1885, at the age of 27, Moser joined Goerz, where he worked on the design of single lenses as landscape objectives and also on the design of Rapid Rectilinear objectives using the new Jena glasses. In this way he produced the Goerz Paraplanat in 1888 and the Lynkeioskop in 1890. He died in 1892 at the early age of 34 and was succeeded by Emil von Höegh.

REFERENCE: Von Rohr's *Theorie und Geschichte*, pages 350–352.

1968

Richard Helmut Naumann

1903 – 1985

Richard Helmut Naumann was born on Christmas Eve 1903, the son of a schoolmaster, in Gross-Röhrsdorf near Dresden. His father was an enthusiastic amateur photographer and he gave Helmut an early liking for photography. He entered the Dresden Technical Institute in 1922, where he concentrated on mathematics and physics and was taught optics from Professor Klughardt and photography from Professor Luther. He obtained a PhD degree with a thesis on color theory in 1928.

He then became scientific adviser to the firm of Emil Busch in Rathenow, where he remained until the end of World War II, in 1945. During this time he designed the first practical zoom lens for 16 mm motion pictures, called the Vario Glaukar; he is shown in the picture holding a Siemens camera equipped with this lens. It covered a range of focal lengths from 25 mm to 80 mm at $f/2.8$. He also wrote several papers and books, including *Das Auge meiner Kamera* in 1937, which has enjoyed several reprintings since.

In 1945 Naumann left Busch and joined Voigtländer in Brunswick. There he taught optics at the Technical Institute and also wrote more books, including *Optik für Konstrukteure* in 1949. He transferred to Rodenstock in Munich in 1954 as head of the Patent and Literary Bureaus, where he remained until his retirement at the age of 70 in 1973. He died in Munich early in 1985.

REFERENCES: Marquis, *Who's Who in Optical Science and Engineering*, page 215, 1985; *Optik* **39,** 475 (1974).

Joseph Max Petzval

1807 – 1891

Joseph Max Petzval was born of Hungarian parents in Spisska-Biela, in the High Tatra mountains of Slovakia. He studied engineering at the University of Budapest, becoming professor of higher mathematics there in 1835, at the age of 28. Two years later he was appointed to a similar position at the University of Vienna. Petzval remained in Vienna until his death.

His major optical work was the design of his famous Portrait lens and his later Orthoskop. In 1845 he quarreled with Voigtländer, who he had entrusted with the manufacture of his two lenses, and turned to Dietzler for help. Unfortunately, Dietzler failed in 1862, after which Petzval abandoned his interest in optics. He married his housekeeper in 1869, but she died four years later. In 1873 he was appointed a member of the Hungarian Academy of Science. Petzval died in 1891, an embittered old man.

REFERENCE: J.M. Eder, *History of Photography*, trans. E. Epstean. Dover, New York, 1978.

1836

G. Simon Plössl

1794 – 1868

It appears that G. Simon Plössl was apprenticed to Voigtländer in 1812 when he was 18, and in 1823 he decided to establish his own company in Vienna. There he made microscope objectives, which he designed himself, and opera glasses. At the 1830 Scientific Congress in Heidelberg he received a prize for the best achromatic microscope. In 1839 he is reported to have made a daguerreotype camera and modified the Chevalier landscape lens. He died in 1868 at the age of 74. In the same year the Optical Society of Vienna named their medal after him.

REFERENCE: J.M. Eder, *History of Photography,* trans. E. Epstean, p. 289. Dover, New York, 1978.

Robert Richter

1886 – 1956

Robert Richter was born in Berlin in 1886. He studied mathematics and physics at various universities and received a PhD degree from Göttingen. He then spent nine years as an optical designer at Voigtländer, from 1914 to 1923, when he joined Goerz. When Goerz was taken over by Zeiss-Ikon in 1926, Richter was the only Goerz optical designer retained by Zeiss. He worked at Zeiss for nearly thirty years until his death early in 1956. He was made chief of Zeiss's photographic division from 1939 to 1945. Most of his designs were aerial camera lenses, including the Topogon, Telikon, Pleogon, and Topar. He also designed the aspheric eyepiece for the Deltar binoculars and the Kipronar $f/1.9$ movie projection lens. In 1933 Richter designed a zoom condenser for a microscope, which anticipated the Pan Cinor design by many years.

REFERENCES: *Photogrammetric Engineering* **22**, 868 (1956); *Optik* **13**, 560 (1956).

L. O. Moritz von Rohr

1868 – 1940

L. O. Moritz von Rohr was born in April 1868 at Lazyn near Inowraclaw, Poland, the son of Louis A. von Rohr, the manager of an estate in Lazyn. He attended the local school and the University of Berlin, where he obtained a PhD degree in 1892. He then became assistant in the Meteorological Institute in Berlin, and in October 1895 he moved to Jena to join the Zeiss Company. A member of several scientific societies, he was also a voluminous writer and is said to have written some 500 books and journal articles. He designed one of the earliest lenses to be equipped with a field flattener.

REFERENCE: *Focal Encylopedia.* Vol. **6,** p. 1323.

Andrew

Thomas

Andrew Ross

1798 – 1859

Thomas Ross

1818 – 1870

Andrew Ross was born in London and attended Christ's Hospital school. He left school at age 14 and was apprenticed to the optician Gilbert, maker of astronomical and surveying instruments, where he eventually became manager of the factory. In 1830 Ross left Gilbert to establish his own company to make microscopes as well as microscope objectives, which were considered the best in the world. On the announcement of the Petzval Portrait lens Ross attempted to make camera lenses, notably a high-aperture portrait lens for the painter Henry Collen, but this was unsuccessful. He then left the photographic side of the business to his son Thomas and his apprentice J. H. Dallmeyer. On his death in 1859 he is said to have left a fortune estimated at £20,000.

On the death of Andrew Ross, management of the company fell to his son Thomas, then 41 years old. He and Dallmeyer evidently decided to separate. In 1864 Ross developed his well-known series of camera lenses known simply as Doublets, based on his father's Collen lens. In 1870 Ross was succeeded as manager by John Stuart, assisted by the engineer F. H. Wenham, who was active in the development of the microscope. Later Stuart was joined by G. A. Richmond and A. A. Smith and by several opticians from Germany including H. L. H. Schroeder, F. G. Kollmorgen, W. F. Bielicke, and J. W. Hasselkus, but most of these left to join other companies or to establish their own, leaving only Hasselkus and Richmond. Ross became Zeiss's London agent in 1890, making many Zeiss lenses under license. After 1900 they developed several types of telephoto lens. In 1897 the company became known as Ross Ltd; it was taken over by Charles Parsons in 1922 and by Barnet-Ensign in 1949, after which it disappeared.

REFERENCES: *Journ. of Phot.* **6,** 49, 52, 234 (1859); *Ibid* **22,** 150 (1875); R. S. Clay, "The Photographic Lens from the Historical Point of View." *Phot J.* **62,** 459 (1922).

ca 1895   1905

ca 1923   ca 1933

Paul Rudolph

1858 – 1935

Paul Rudolph was born in November 1858 at Kahla, in Thuringia, and attended universities in Munich, Leipzig, and Jena, where he graduated in 1884. He originally intended to be a teacher in mathematics and physics, but in 1886 he was invited to join Zeiss as a mathematician and assistant to Abbe. He organized Zeiss's photographic department in 1889. He started with the design of the Apochromatic Triplet, with E. Abbe, but it was no better than other lenses and was withdrawn. He then went on to design the Anastigmat series, which sold well in Germany and other countries for many years. In 1896 he produced the Planar, in which he introduced the idea of a buried surface to correct the chromatic aberration. The Unar followed in 1899 and then the famous Tessar in 1902. The latter design earned him the Progress Medal of the Royal Photographic Society in 1905.

Because at that time designers were paid a royalty on their inventions, Rudolph became a rich man and in 1911 he retired to become a country gentleman. Unfortunately, he lost everything in the inflation that followed World War I and had to go back to work at the age of 62. He returned briefly to Zeiss, but as he preferred to work in a small company, he joined Hugo Meyer in Görlitz, where he designed a series of Plasmat lenses of several types. He retired for the second time in 1933 and died two years later.

REFERENCES: *Phot. J.* **45,** 91 (Mar. 1905); obituary and an outline of his life by von Rohr, *Phot. J.* **75,** 357 (1935).

Willy E. Schade

1889 – 1973

Willy E. Schade was born in Braunschweig in Germany, where he worked briefly at the Voigtländer plant. He served in the German army on the eastern front during World War I and was wounded. After the war he moved to Berlin, where he worked at Goerz under F. Weidert, but he was let go when Goerz was absorbed into the Zeiss-Ikon combine in 1926. At that time he decided to emigrate to the United States and obtained a position as lens designer at Ilex in Rochester. Six years later he transferred to Kodak, where he became the designer of many of their best lenses. He retired in 1958 and returned to his home town of Braunschweig, where he died in 1973.

Willy Schade was an old-fashioned designer and remained faithful to the book of six-figure logarithms until his retirement, although he had no objection to his assistants using any computing aids they desired. He used to wear out a copy of Bremiker's logarithms every year!

Bernhard Voldemar Schmidt

1879 – 1935

Bernhard Schmidt (the name was originally Matts) was born in Nargen, Estonia, and remained an Estonian citizen all his life, although he lived mostly in Germany. A boyhood experiment with gunpowder cost him his right arm. After a few early jobs in Estonia he moved to Mittweida near Jena, where he set up a workshop, and by 1903 he was making excellent parabolic mirrors of eight-inches diameter. During World War I he was interned briefly as an enemy alien, but he was allowed to work on periscopes for the army. He delivered his first mirror to Bergdorf Observatory in 1918.

After the war he returned to Estonia, but in 1926 he was given the use of a basement workshop at Bergdorf. In 1932 he made his first telescope with a spherical mirror and an aspheric corrector plate, although he did not reveal how it was made. He died in 1935 with the secret still intact.

REFERENCE: *Sky and Telescope* **15,** cover and page 4 (Nov. 1955). H.C. King, *History of the Telescope,* page 356. Sky Publishing Co., Cambridge MA, 1955.

Joseph Schneider Sr.

d. 1933

In the late 1800s, Joseph Schneider Sr. operated a brewery in Springfield, Illinois. When he foresaw the coming of prohibition, he returned to his home in the Rhineland, where he cultivated grapes and wine, eventually becoming vintner to the Czar. His son Joseph August studied at Frankfurt University, where he became interested in optics, so in 1913 Joseph Sr. sold his vineyards and together with his son founded the Joseph Schneider Optical Works in Kreuznach.

They made their first $f/4.5$ Xenar lens in 1919 and their millionth lens was made in 1936, by which time they had over five hundred workers at their factories in Kreuznach and at Isco in Göttingen. After Joseph August's death in 1950 the firm was headed by his widow and their son Hans Joseph. By 1954 they had over 1,000 workers, making a wide range of lenses for all purposes, by which time they had made some eight million objectives. In 1963 the company celebrated its fiftieth anniversary.

REFERENCE: *Camera News of West Germany,* September 1963.

1883

ca 1920

Friedrich Otto Schott

1851 – 1935

Friedrich Otto Schott was born at Witten, near Essen, where his father was manager of a plate glass factory. He studied chemistry at Aachen, Wurzburg, and Leipzig, where he graduated after writing a thesis on "Defects in the Manufacture of Window Glass." In 1879 Schott sent some samples of a new lithium glass to Abbe for review, but they were not what Abbe wanted. Schott then turned to glasses containing boron and phosphorus, which were much more promising, so in 1881 Abbe and Schott started a collaboration by mail with a number of small experimental melts. These were so promising that Abbe finally invited Schott to move to Jena and join him in establishing a glass factory there. By 1884 the factory was running, and two years later they published their first catalog containing 44 glasses, many of which were entirely new types. Schott died in 1935 at the age of 84.

After the expropriation of the Jena Glass Works in 1952, the firm moved to Mainz in the western zone of Germany. The demand for Schott glass in America was so great that in 1967 they established a branch factory at Durea, Pennsylvania. Their current catalog is a veritable textbook on the properties of optical glass.

Heinrich Ludwig Hugo Schroeder

1834 – 1903

Hugo Schroeder served his apprenticeship with M. Meyerstein in the university workshop at Göttingen, and he worked with J. B. Listing in that city. He then moved to Hamburg where he made telescopes. In the late 1870s, he moved to Oberursel near Frankfurt, but his business did not do well and he was happy to accept an invitation from John Stuart, manager of Ross and Co. in England, to join that company in 1882 and to take charge of their scientific work. In 1888 he patented the Concentric lens, one of the first designs to make use of the new barium crown glasses recently developed by Abbe and Schott at Jena. In 1894 he spent a year in the United States working with the Manhattan Optical Company. He returned to England in 1895 and lived a life of retirement.

REFERENCE: *Brit. J. Phot. Almanac,* 1903, page 688.

Charles Piazzi Smyth

1819 – 1900

C. Piazzi Smyth was born in Naples, Italy, of English parents. In 1825 his parents returned to England and settled in Bedford, where Charles attended school. From 1835 to 1845 he worked at the Cape Observatory as assistant to Maclear. In 1845 he was appointed director of the Edinburgh Observatory and Astronomer Royal for Scotland at the age of 26.

In the early 1960s he became obsessed with the idea that the Great Pyramid in Egypt had some occult significance, and in 1865 he went to Egypt to study the subject. He made a small camera with a Petzval Portrait lens, but the astigmatism worried him. In 1874 he proposed the use of a negative field flattener lens close to the image. His papers on the Great Pyramid were refused publication by the Royal Society, so he resigned in disgust. He retired from Edinburgh in 1888 and died near Ripon in Yorkshire in February 1900.

REFERENCES: *Dictionary of Scientific Biography* **12,** 498 (1975); *History of Photography* **3,** 332 (1979) and **4,** 290 (1980); *Pop. Astronomy* **8,** 384 (1900).

Carl August von Steinheil

1801 – 1870

Carl August Steinheil was born in Rappoltsweiler in Alsace, the son of Karl Philipp Steinheil, administrator of the estates of the prince who became Maximillian I of Bavaria. In 1823 he studied science and astronomy under Gauss at Göttingen and Bessel at Konigsberg. He obtained a PhD at the latter university under Bessel in 1825. In 1832 he became professor of mathematics and physics at Munich. When the daguerreotype process was announced, he, along with his colleague Franz Kobell, made a miniature camera that produced images 8 mm × 11 mm on polished silver coins.

Steinheil was an early pioneer in electrical work. In 1849 he was invited to organize the Bavarian telegraph system and two years later a similar system in Switzerland. In 1854 with his son Adolph he founded the Steinheil Optical Institute in Munich, where he died in 1870. In 1856 he collaborated with Foucault in the deposition of silver on telescope mirrors.

REFERENCE: *Dictionary of Scientific Biography* **XIII,** p. 22, 1975.

Hugo Adolph Steinheil

1832 – 1893

August's son Adolph was strongly inclined toward optics and astronomy. He studied at Munich and Augsburg, and accompanied his father to Vienna in 1850 at the age of 18 while the latter was organizing the telegraph system there. Adolph returned to Munich in 1852 to devote himself entirely to optics. He took a prominent part in the establishment of the Steinheil Optical Institute there in May 1855. Working with his friend L. P. von Seidel (1821 – 1896), Adolph designed the Periskop in 1865 and the Aplanat in 1866. There was a great argument as to the priority of the invention of the Aplanat and Dallmeyer's Rapid Rectilinear, and it appeared that Steinheil had priority, but only by a few weeks.

In 1866 Adolph purchased his father's interest in the Optical Institute, which then became C. A. Steinheil Sohne. Adolph went on to design many other lenses, including the Group and Portrait Antiplanets in 1881. He collaborated with Ernst Voit in writing a book on lens design in 1891, two years before his death.

REFERENCE: J. M. Eder, *History of Photography*, trans. E. Epstean, p. 403. Dover, New York, 1978.

Rudolph Steinheil

1865 – 1930

The third member of the famous Steinheil family was Adolph's son Rudolph, who was an accomplished lens designer. At age 28 he became head of the firm, and the male line ended at his death in 1930. The firm then became a stock company owned by his five daughters. It was acquired by Elgeet in 1962.

Rudolph Steinheil began by extending his father's interest in the Antiplanet principle by the Rapid Antiplanet of 1893. By use of the new Jena glasses he developed the rear of the Antiplanet to make the excellent Orthostigmat. He went on to design other lenses, including the Unofocal of 1901. He also made several large telescopes for German observatories. In 1910 he collaborated with Goerz in the establishment of the Sendlinger Glassworks in Berlin.

REFERENCES: *Brit. J. Phot.* **77,** 98 (1930); *Photo. Korresp.* **66,** 107 (1930).

THE LATE THOMAS SUTTON, B.A. (CANTAB.)

Thomas Sutton

1819 – 1875

Thomas Sutton was born in Kensington and attended Cambridge University, where he graduated twenty-seventh. Wrangler in 1846. In 1847 he moved to Jersey where he pursued optical and photographic studies and founded the magazine *Photographic Notes* in 1856. In September 1859 he patented the unusual water lens called the Panoramic. He realized that symmetry automatically removes distortion and worked with R. H. Bow on the theory of distortion. In 1867 he lived briefly at Redon in Brittany. He returned to England in December 1874 and died at Pwllheli the following year.

REFERENCES: M. von Rohr, *Theorie und Geschichte* p. 194; obituary and portrait, *Brit. J. Phot.* **22,** page 210 (1875); [reprinted May 2, 1975].

Harold Dennis Taylor

1862 – 1943

Harold Dennis Taylor was born in Huddersfield, England, and attended St. Peter's School, York. He joined T. Cooke and Sons, telescope makers, in York, and eventually became their optical manager. In 1892 he patented a three-element apochromatic telescope objective (Brit. Pat. 17,994/92), and in 1893 he designed his famous Triplet photographic objective (Brit. Pats. 1991/93 and 22,607/93). But as Cooke did not wish to make photographic objectives, he offered the design to Taylor, Taylor and Hobson [no relation] in Leicester, who undertook to manufacture it. In 1896 Taylor noticed that a tarnished lens transmits more light than a freshly polished one, thus laying the foundation for antireflection coating. In 1906 Taylor wrote his famous book *A System of Applied Optics*. He was awarded the Progress Medal of the Royal Photographic Society in 1935.

REFERENCE: *Nature* **151,** 442 (1943).

Johann Christoph

Johann Friedrich   1832

Johann Christoph Voigtländer

1732 – 1797

Johann Friedrich Voigtländer

1779 – 1859

Johann Christoph Voigtländer (the name was originally Vogtländer) was born in Blankenburg in the Harz Mountains and in 1756 founded a small company to make fine mechanical instruments. He was assisted by his three sons, Wilhelm (1768 – 1828), Siegmund (1770 – 1822), and Johann Friedrich (1779 – 1859), the youngest of whom carried on the business after their father's death in 1797.

As a young man, Johann Friedrich Voigtländer went to England to learn the trade of optician and after his return to Vienna started to make lenses in 1808. It is said that he introduced Wollaston's meniscus spectacles into Germany and Austria and started to make opera glasses in 1823. One of his apprentices was Simon Plössl (1794 – 1868), who left in 1823 to found his own company. Plössl and Voigtländer were considered to be the finest opticians in Vienna. He retired in 1837 at the age of 58.

REFERENCES: H. Harting, *Zūr Geschichte der Familie Voigtländer*. Central-Zeitung für Optik und Mechanik, 1924 – 1925 (published as a separate booklet by the Voigtländer Company). There are also many references to Voigtländer in Eder's *History of Photography*, Dover, New York, 1978.

Peter Wilhelm Friedrich   1840

Friedrich Ritter (von)   1915

Peter Wilhelm Friedrich Voigtländer

1812 – 1878

Friedrich Ritter von Voigtländer

1846 – 1924

On Johann Friedrich's retirement in 1837, management of the firm was taken over by his son Peter Wilhelm Friedrich, then just 25 years old. He was a very fine optician, and it was to him that Petzval entrusted the making of his Portrait Lens in 1840, which was installed in a simple conical metal camera. Young Voigtländer established a branch factory in Brunswick in 1849 and closed the Vienna factory in 1866. It is reported that in 1860 he was making Petzval Portrait lenses as large as 6 inches in diameter; by 1862 he had made some 60,000 Petzval lenses. He was ennobled by the King of Austria in 1866, which entitled him to use the prefix von before his name. He retired in 1876 at the age of 64.

The last of the four Voigtländers was Friedrich Ritter. He managed the company from 1876 until its conversion into an AG in 1898; he then became Chairman of the Board, with his five daughters as owners.

In 1870 Friedrich worked with his half-brother Hans Zinke-Sommer on the design of an $f/2.37$ Portrait lens, and he produced an all-cemented design in 1878. He was a fine designer, working on the Euryskop in 1888 and numerous other lenses. His two assistants, David Kaempfer and Hugo Scheffler, produced the famous Collinear lens in 1892. Other Voigtländer designers were Hans Harting and Robert Richter. The company was taken over by Zeiss-Ikon in 1965.

REFERENCES: H. Harting, "Zür Geschichte der Familie Voigtländer." Central-Zeitung für Optik und Mechanik, 1924 – 1925 (published as a separate booklet by the Voigtländer Company). There are also many references to Voigtländer in Eder's *History of Photography*.

Joseph Bailey Walker

1892 – 1985

Joseph Walker was born of wealthy parents in Denver, Colorado, and his first technical interest was in early wireless; he later worked briefly with Lee DeForest in that field. However, by the age of 22 he had become fascinated with motion picture production, and he eventually became cameraman for Frank Capra. During his long career he photographed some 160 feature productions between 1919 and 1952, including such well-known pictures as "It Happened One Night" (1934) and "Lost Horizon" (1937).

In 1932 he assembled a primitive zoom lens by mounting a large negative viewfinder lens in front of an ordinary motion-picture camera, and he observed that he could vary the image size by moving the negative lens closer to or further from the camera. After World War II this basic design was incorporated into the Electrozoom lens for television cameras, which Walker patented and manufactured. Eventually this idea was developed into the two-component zoom lens for SLR cameras because of its extreme wide-angle capacity. Walker died in Las Vegas at the advanced age of 92, after receiving many honors from the motion-picture industry.

REFERENCE: J. B. Walker and J. Walker, *The Light on Her Face*. ASC Press, Hollywood, 1984. Obituary in *J.S.M.P.T.E.* **94**, 1210 (Nov. 1985).

1959

Ernst Wandersleb

1879 – 1963

Ernst Wandersleb was born in Gotha in April 1879. His father was a primary school teacher and also taught music. Ernst obtained his doctorate in physics at Jena University under A. Winkelmann in 1900.

In 1901 he joined Zeiss and served first as assistant to von Rohr and then to Paul Rudolph. In 1906 he married Emmy, sister of his colleague Otto Eppenstein. When Rudolph retired in 1911 Wandersleb became his successor as manager of the photographic lens department. Among other designs, he raised the aperture of the Tessar to $f/3.5$. He was fond of sports and took up ballooning at about that time.

Unfortunately, with the coming of National Socialism in 1938, he and his family were declared *halbarisch* and suffered badly. His children were not allowed to pursue their professional careers and his wife Emmy, a Jew, was sent to the camp at Buchenwald. Wandersleb expressed great criticism of the Nazi regime and was not allowed to remain at Zeiss. After the war he was rehired in 1945 and finally retired in 1957. Both he and his wife died in 1963.

REFERENCES: *Zeit. für Instkde* **27**, 33, 75 (1907); *Optik* **20**, 512 (1963) and **22**, 577 (1965).

Arthur Warmisham

1891 – 1962

Arthur Warmisham was born in January 1891 and studied at Manchester University where he obtained an MSc degree under Rutherford. He joined Taylor-Hobson in 1912 at the age of 21 and became Director of Research there from 1922 until his retirement in 1955 after 43 years of service. He was made a Fellow of the Institute of Physics in 1938. In the early 1930s he spent some time in Chicago with the Bell & Howell Company, the American agents for Taylor-Hobson lenses, after William Taylor had sold his shares to Bell & Howell. (These shares were acquired by J. Arthur Rank in 1946.) Warmisham died in November 1962. He owned many patents for his numerous novel types of photographic lens, which included the earliest practical zoom lens, the Varo, of 1931.

REFERENCE: D. F. Horne, *Optical Instruments and their Applications*, p. 32. Hilger, Bristol, 1980.

William Hyde Wollaston

1766 – 1828

William H. Wollaston was born at East Dereham, Norfolk, in 1766. He was educated at Charterhouse School and Caius College, Cambridge. He received a medical degree in 1793 and practiced for a few years. He was elected to the Royal Society in 1793 and became secretary to the society from 1804 until 1816. He died in London at the age of 62.

His scientific work was mainly in chemistry, and he made a small fortune from his discovery of the practical working of platinum. He worked also with several of the other rare elements. In optics he was the first to observe the dark lines in the solar spectrum in 1802. He developed the camera lucida in 1807, chiefly as an aid to drawing objects seen through the microscope. He suggested the use of meniscus-shaped lenses for spectacles in 1804 and extended that idea to the lens on a camera obscura in 1812.

REFERENCES: *Encyclopedia Britannica* **23**, 700 (1961); *Phil. Mag.* **41**, 124 (1813).

Andrew

John Charles

Andrew Wollensak

1862 – 1936

John Charles Wollensak

1864 – 1933

Andrew (or Andreas) Wollensak was born in Wiesbaden, Germany, and came to the United States in 1882, where he entered into the employ of Bausch & Lomb as a machinist. After 1890 he worked on the design of shutters and later became a foreman. In July 1899 he decided to establish his own shutter company with his brother John, who was also born at Wiesbaden and moved to Dunkirk, New York, in 1886. They were assisted financially by Stephen Rauber, a retired brewer. The name of the company was Rauber and Wollensak until 1902, when Rauber retired to become a coal merchant.

The company grew slowly and after several changes in location finally settled on Hudson Avenue in Rochester in 1924. They manufactured a wide range of shutters, including the famous Optimo of 1909. They added lenses to their line in 1902 and acquired the assets of the Rochester Lens Company in 1905. By 1946 the number of employees reached 1,200. Among other ventures they manufactured the Fastax high-speed camera for the Bell Laboratories; it was acquired by the Redlake Corporation in 1968. The Wollensak Company was acquired by Revere in 1953 and by 3M in 1960. Wollensak purchased the Polan Company in Huntington, West Virginia, and moved their headquarters there for a short time. In recent years production dropped and they ceased operations in 1972.

REFERENCE: *Rochester in History,* p. 189 (1922).

Joseph Zentmayer

1826 – 1888

Joseph Zentmayer was born in Mannheim, Germany, where he learned the trade of an instrument maker. He came to the United States in 1848 at the age of 22 and for the next five years worked with various instrument makers in Baltimore, Washington, and Philadelphia. In 1853 he set up his own workshop at Eighth and Chestnut Streets in Philadelphia, where he began to make microscope stands and objectives. It is recorded that he placed his lathe in the front of his shop so that he could talk to customers while working on his microscopes. He apears to have been a most lovable human being.

He developed his only photographic objective in 1866. Having read a description of the Harrison Globe lens, it occurred to him that it might be possible to make a similar lens using simple elements. This he did, the focal lengths of the elements forming a series, each having a focal length equal to 1½ times the next in the series, so that by combining two adjacent lenses about a central stop, he could make a series of double lenses with increasing focal length.

REFERENCES: *Proc. Am. Micros. Soc.* **14,** 161 (1892); *J. Franklin Inst.* **82,** 63 (1866) and **126,** 483 (1888).

Walther O. W. Zschokke

1870–1951

Walther Zschokke, a Swiss optical designer, was employed first by C. A. Steinheil in Germany and then moved to C. P. Goerz in 1902 as successor to Emil von Höegh, who had resigned for reasons of ill health. He collaborated with Franz Urban in the design of several excellent Goerz lenses, including the Alethar, Pantar, Dogmar, and the long-lived Artar, which became the most used process lens anywhere. In 1920, after World War I, Zschokke returned to Switzerland, where he established an optical department at Kern in Aarau. For some reason he left Kern in 1925 and set up as an optometrist. He died in 1951 at the age of 81.

REFERENCE: *Image*, George Eastman House, **28,** 17 (March 1985).

APPENDIX

Glossary of Optical Terms

The brief definitions given here represent only the basic concepts in each case. Lens designers require much more detailed and specific definitions, which are beyond the scope of this book.

Abbe V-number. The Abbe, or V-number, of a glass is the reciprocal of the dispersive power. Hence $V = (n_d - 1)/(n_F - n_C)$.

Aberration. A property of a lens causing a distortion or loss of definition in an image. There are seven basic types of aberration, known respectively as spherical, coma, astigmatism, field curvature, distortion, chromatic, and lateral color.

Achromatic lens. A lens in which two different wavelengths of light come to a focus at the same distance from the lens.

Anamorphic lens. A lens in which the image magnification varies with the orientation about the lens axis, reaching maximum and minimum values in a specific pair of orientations, generally vertical and horizontal.

Anastigmat lens. A lens in which the stigmatic image at one point in the field falls in the same plane as the axial image. The name, capitalized, was used by Zeiss to refer to a particular type of construction.

Angular field. *See* field of view.

Antireflection coating. *See* lens coating.

Aplanat. An old term for a lens corrected for spherical aberration. Today this term also implies correction for coma.

Apochromat. A lens in which three wavelengths instead of the usual two meet at a common focus.

Aspheric surface. A surface of revolution about the lens axis, the longitudinal section of which is not a circle. It may be a conic section or a more general asphere having several orders of asphericity.

Astigmatism. An aberration in which light from an extraaxial object point forms a pair of focal lines after passing through the lens. One of these focal lines is tangential to the field of view while the other is radial to it. The magnitude of the astigmatism is expressed by the longitudinal separation between these two focal lines. If the tangential line is closer to the lens than the radial line the astigmatism is said to be negative or undercorrected. *See also* field curvature.

Axial object point. A point object lying on the lens axis.

Axis. The line joining the centers of curvature of all the surfaces in a lens. If one or more of the centers of curvature lie to one side of the axis the lens is said to be decentered.

Bending. The name given to the process of changing the shape of a lens without altering its focal length.

Buried surface. A cemented interface inserted between two lens elements made of glasses having the same refractive index but different dispersion.

Catoptric and catadioptric systems. Optical imaging systems involving mirrors only (catoptric) or a combination of lenses and mirrors (catadioptric).

Chief ray. *See* principal ray.

Chromatic aberration. An aberration in which images in different wavelengths (colors) fall at different distances from a lens. In a simple positive lens the blue image falls closer to the lens than the red image; this is said to be negative or undercorrected chromatic aberration.

Coated lens. A lens that has been given an antireflection coating on its polished surfaces. *See also* lens coating.

Collective and dispersive surfaces. Lens surfaces that act to collect rays together or spread them apart, respectively.

Coma. An aberration causing a radial flare from the image of an extraaxial object point.

Component. A single lens element, or an assembly of elements that act together as a group. In a zoom lens some components move while others remain stationary.

Crown glass.   An optical glass having a V-number greater than 50.

CRT.   An abbreviation for cathode ray tube.

Decentered lens.   *See* axis.

Degree of freedom.   To the lens designer a degree of freedom is some property of a lens that he is free to alter. Examples are the radii of curvature of the surfaces, the thicknesses of the elements, airspaces between the elements, and the types of glass. If there are aspheric surfaces in a lens, the aspheric coefficients are also degrees of freedom.

Depth of field.   The finite range of object distances within and beyond the focused plane that appear acceptably sharp in a photograph. Depth of field depends mainly on the distance of the observer from the picture; it also varies with the linear aperture of the camera lens and the distance of the original object from the camera.

Dialyte.   A lens consisting of a separated pair of positive and negative elements.

Diffraction.   A narrow-angle scattering of light at an edge, such as at the blades of an iris diaphragm. This often causes streaks of light to radiate from any bright light in the scene, each streak being perpendicular to the edge causing it. If the aperture of a lens is very small, say on the order of one millimeter, diffraction may cause a general loss of definition in a photograph.

Diopter.   A measure of lens power equal to the reciprical of the focal length in meters.

Dispersion.   The dispersion of a transparent medium is the difference between the refractive indices for a specific pair of wavelengths. Generally the chosen wavelengths are the red and blue C and F Fraunhofer lines. The dispersion is then given by $(n_F - n_C)$.

Dispersive power.   The dispersive power of a transparent medium is equal to the ratio of the dispersion to the mean refractive index minus 1.0.

Distortion.   An aberration in which the image magnification is not constant over the whole image. If the magnification at the edge is greater than that on the axis the distortion is said to be pincushion or positive; if it is smaller at the edge the distortion is said to be barrel or negative.

Doublet lens.   This term has two possible meanings: (a) a pair of lens elements cemented together or (b) a system consisting of two separated components with a central airspace containing the stop.

Element.   A single piece of glass having two polished surfaces. The axis, or line joining the centers of curvature of the surfaces, passes through the middle of the piece.

Extra-axial object point.   An object point lying to one side of the lens axis.

Field curvature.   A measure of the departure of the image of a plane object from being also a plane. Specifically, the field curvature of a lens is the locus of the tangential focal lines in the images of a series of object points lying on a flat plane perpendicular to the lens axis. There is also a second image field formed by the locus of the radial focal lines across the image. The separation between the two fields is a measure of the astigmatism.

Field flattener.   A negative lens placed at or close to an image to reduce the Petzval sum without affecting the focal length or any of the other aberrations.

Field of view (FOV).   The extent of the angular area in the object space for which the lens forms an acceptably sharp image.

Fish-eye lens.   A lens of the reversed telephoto type having so much barrel distortion that an entire hemisphere is imaged as a finite circle on the film.

Flare spot.   A faint out-of-focus image of the lens diaphragm formed in the middle of the picture and superimposed on the desired image. It is caused by interreflections of light between the lens surfaces in the rear component of the lens, between the iris and the image.

Flint glass.   An optical glass having a V-number less than 50.

Floating lens.   A recent development in which one or more of the lens elements are moved independently during focusing, in such a way that the aberrations remain well corrected throughout the focusing range.

F-number.   The F-number of a lens is the numerical value of the ratio of the focal length to the linear diameter of the lens aperture.

Focal length.   The focal length of a lens is that property which determines the size of the image of a distant object. Specifically, focal length is defined as the limiting value of the image size divided by the angular extent of the object, both quantities being measured out from the lens axis. Thus

$$f = \lim_{\theta \to 0} \left(\frac{h'}{\tan \theta} \right)$$

where h' is the image height and θ is the angular subtense of the object.

Focus.   As a noun, the focus is the plane of sharpest imagery. As a verb, to focus an image is to set the receiving surface at the point of sharpest imagery. The principal focus of a lens is the point of sharpest imagery for a very distant object.

Focus shift. A phenomenon caused by excessive zonal aberration, in which the plane of sharpest definition moves along the axis when the lens is stopped down to a smaller aperture.

Fraunhofer lines. A series of dark lines in the solar spectrum, used by Fraunhofer as fiducial marks to indicate specific points in the spectrum.

F-theta lens. A lens having so much barrel distortion that the image height is given by the product of the focal length and the field angle in radians.

Galilean telescope. A small telescope consisting of a positive objective and a negative eyepiece. It forms an erect image and covers only a moderate angular field at a low magnifying power.

Gauss objective. A type of photographic lens consisting of a pair of components about a central stop, each component containing a meniscus-shaped negative unit close to the stop and a positive unit outside.

Ghost image. An unwanted image, or a series of images, of a bright light located in or close outside the scene, caused by interreflections between the various lens surfaces. The line of images always joins the light source to the center of the picture. Ghost images can be drastically reduced by appropriate lens coating.

Gradient index (GRIN) material. A form of transparent material in which the refractive index varies in a continuous manner throughout the piece, either radially or longitudinally.

Image. An apparent reproduction of an object formed by a lens or mirror or by other means.

Inverted image. An image that is rotated through 180° about the lens axis as compared with the orientation of the original object.

Iris diaphragm. A mechanical means for varying the size of a lens aperture.

Landscape lens. A simple type of lens covering a semi-angular field of about 20° at a low aperture, suitable for landscape photography.

Lateral color. An aberration in which the size of an image varies with the wavelength (color) of the light. If the blue image is larger than the red image the lateral color is said to be positive.

Left-handed image. An image such as that seen in a mirror, in which the viewer holding something in his right hand sees the image as holding it in his left hand.

Lens. Either a single lens element or a complete optical system.

Lens aperture.   The entering diameter of the largest axial beam of light that can pass through the lens.

Lens coating.   A process for reducing the reflectivity of a glass surface by depositing one or more thin layers of a low-index material on it.

Lens hood.   An empty tubular extension of a lens mount in the object space to prevent unwanted light from a bright source such as the sun from entering the lens.

Macro lens.   A lens which is well corrected for use over a wide range of object distances.

Magnification.   The lateral dimension of an image divided by the corresponding dimension of the object.

New achromat.   A cemented doublet consisting of a positive element of high-index crown glass attached to a negative element of low-index flint glass.

Nodal points.   A pair of conjugate axial points in a lens such that a paraxial ray directed towards the first leaves as from the second at the same slope as it enters. The distance from the second nodal point to the principal focus is equal to the focal length of the lens.

Objective.   The lens in a camera, telescope, microscope, or other optical system that lies closest to the object being observed.

Optical glass.   Glass intended specifically for optical purposes. It has a known and carefully controlled refractive index and dispersion, with excellent homogeniety, color, freedom from bubbles and striations, and with a high degree of chemical stability.

Optician.   An individual or a company that manufactures lenses or other optical devices.

Paraxial ray.   A ray that throughout its length lies infinitely close to the optical axis.

Periscopic lens.   A lens consisting of a pair of identical or similar thin meniscus elements mounted symmetrically about a central stop, the concave sides of the lenses facing the stop.

Petzval sum.   A certain sum $\Sigma(n' - n)/nn'r$ in a lens. The magnitude of this sum determines the basic field curvature of the lens, and also the magnitude of the residual astigmatism after the tangential field has been flattened.

Photography.   The formation of images by optical, chemical, electronic, or xerographic means.

Portrait lens.   Any lens having a sufficiently high relative aperture for convenient use in portrait photography. Specifically, the Petzval Portrait lens of 1840.

Positive and negative lenses.   A positive lens is a single element that is thicker in the center than at the edge. A negative lens is thinner at the center than at the edge.

Power of a lens.   The power of a lens is equal to the reciprocal of the focal length. *See also* diopter.

Principal ray.   The middle ray of a beam of light passing through a lens. It is becoming usual to distinguish between the principal ray and the chief ray. The principal ray is now defined as the ray in an oblique beam that passes through the center of the stop, while the chief ray is the ray that enters the lens midway between the highest and the lowest rays that can pass through it. In the absence of vignetting, the principal and chief rays are identical.

Rapid Rectilinear lens (or Aplanat).   A type of symmetrical photographic lens in which each half consists of a cemented doublet with all three surfaces concave towards the central stop.

Refraction, law of.   When a ray of light is refracted (bent) at the surface of separation between two adjacent materials having different refractive indices n and n', the angles I and I' between the ray and the normal at the point of incidence are related by Snell's law:

$$n \sin I = n' \sin I'$$

Refractive index.   The refractive index of a transparent medium is equal to the ratio of the velocity of light in air to its velocity in the medium.

Relative aperture.   The ratio of the aperture of a lens to its focal length. Relative aperture can be expressed as a ratio, e.g. $1 : 4.5$, or as a fraction $f/4.5$. In either case the figure 4.5 is known as the F-number of the lens.

Retrofocus lens.   A lens of the reversed telephoto type manufactured by the firm of Angénieux in France. It is becoming a generic term for any lens of the reversed telephoto type.

Reversed telephoto lens.   A type of photographic objective consisting of a negative component in front and a positive component in the rear. The back focus, or distance from the rear lens surface to the image plane, is equal to or greater than the focal length.

Sagittal field.   The locus of the radial focal lines in the images of a series of object points lying in a plane perpendicular to the lens axis. *See also* field curvature.

Secondary spectrum.   In an achromatic lens, the secondary spectrum is the departure of other wavelengths from the two that have been united at a common focus.

Sky lens.   *See* fish-eye lens.

Snell's law.   *See* refraction, law of.

Spherical aberration.   A blurring of an image resulting from the use of spherical surfaces in a lens. In a single positive element the marginal rays cross the axis closer to the lens than the central rays, resulting in negative or undercorrected spherical aberration.

Spherochromatism.   The variation of spherical aberration with the wavelength of the light.

Stigmatic lens.   A lens in which the astigmatism has been corrected at one particular obliquity. This name was used, with a capital S, for a series of photographic objectives manufactured by Dallmeyer.

Stop.   A mechanical aperture, often of variable diameter, that limits the size of the beams of light transmitted by a lens.

Symmetrical lens.   A photographic objective in which the front and rear components are identical but one half is turned end-for-end relative to the other half. The two halves are equidistant from a central stop.

Tangential field.   The locus of the tangential focal lines in the images of a series of object points lying in a plane perpendicular to the lens axis.

Telenegative attachment.   A negative lens unit inserted between the lens and film of a camera to increase the focal length.

Telephoto lens.   A type of photographic objective consisting of a positive front component and a negative rear component, such that the focal length is greater than the total length from the front vertex to the image plane. The ratio of the total length to the focal length is known as the telephoto ratio.

Triplet lens.   This term may have two meanings: (a) a single unit consisting of three lens elements cemented together or (b) an objective containing three airspaced elements in order ($+ - +$).

Varifocal lens.   A lens provided with means for varying the focal length in a continuous manner by a longitudinal movement of some parts of the system relative to other fixed components.

Vignetting.   A trimming of oblique beams passing through a lens by the rims of the various optical or mechanical elements in the system.

V-number.   *See* Abbe.

Zonal aberration.   After the marginal and central rays through a lens have been united at a common focus, the rays through other lens zones do not in general pass through the same focus. The difference is known as zonal aberration.

Zoom lens.   A varifocal lens in which the position of the image is held fixed by optical or mechanical means.

LENS NAMES INDEX

The following names of lenses are referenced in the text; they have all been developed since about 1895. Earlier classical pre-anastigmats and some unusual and special lenses are included in the regular index and are not in this list.

INDEX

This index contains topics and names of persons. The names of recent lenses are listed beginning on page 323. The names of companies are not indexed.